Critical [Acclaim for]
Robert S[ullivan]

"[Sullivan] is brilliant at capturing the moods and moments of an American family road trip."

—Bob Minzenheimer, *USA Today*, on *Cross Country*

"Immensely lively, enjoyable, witty, and yes, appealing."

—Phillip Lopate, *Washington Post*, on *Rats*

"A book that is at once enthralling, fair-minded, and very funny."

—Jonathan Raban, *New York Review of Books*, on *A Whale Hunt*

"Provocative, audacious. . . . By looking observantly, without trite moralizing, at the natural world . . . this book suggests a challenging new model for how we ought to pay attention."

—Robert Pinsky, *New York Times Book Review*, on *The Meadowlands*

ALSO BY ROBERT SULLIVAN

Cross Country

How Not to Get Rich; Or Why Being Bad Off Isn't So Bad

Rats

A Whale Hunt

The Meadowlands

THE THOREAU YOU DON'T KNOW

The Father of Nature Writers on the Importance of Cities, Finance, and Fooling Around

ROBERT SULLIVAN

HARPER PERENNIAL

NEW YORK • LONDON • TORONTO • SYDNEY • NEW DELHI • AUCKLAND

HARPER PERENNIAL

A hardcover edition of this book was published in 2009 by
Collins, an imprint of HarperCollins Publishers.

HarperCollins books may be purchased for educational, business, or sales promotional use. For information, please write: Special Markets Department, HarperCollins Publishers, 10 East 53rd Street, New York, NY 10022.

FIRST HARPER PERENNIAL EDITION PUBLISHED 2011.

The Library of Congress has catalogued
the hardcover edition as follows:

Sullivan, Robert
The Thoreau you don't know : what the prophet of environmentalism really meant / Robert Sullivan.—1st ed.
p. cm.
Includes index.
ISBN 978-0-06-171031-5
1. Thoreau, Henry David, 1817–1862. 2. Authors, American—19th century—Biography. 3. Naturalists—United States—Biography. I. Title.

PS3053.S86 2009
818'.309—dc22
[B]

2008034495

ISBN 978-0-06-171032-2 (pbk.)

11 12 13 14 15 OV/BVG 10 9 8 7 6 5 4 3 2 1

For Suzanne

Contents

THE THOREAU YOU DON'T KNOW

Chapter 1

THE THOREAU YOU DON'T KNOW

I'D LIKE TO INTRODUCE the Thoreau you don't know, or don't necessarily know, or know but perhaps never hear people talking about when people talk about Thoreau. People talk a lot about Thoreau in America—they reference him in these days of ecological awareness, in these green times, in times when, as people all along the political spectrum agree, we care about the earth, the wilderness, what's wild. But when we talk about Thoreau, we talk about a particular Thoreau who I would suggest has more to do with us than Thoreau. The Thoreau we already know is, for instance, not here. He's out, away, off in the woods most likely, on the shore of that lonely little pond or ascending a faraway mountain. He's busy getting more in touch with the natural world. He's not in town with us, that's for sure. The Thoreau we know doesn't really go to town. He's managed to separate himself from the hustle, the rat race—or at least to *say* that he has. People wonder, after all. They always have. People don't completely trust this Thoreau. They wonder if he's coming clean, if he really does spend all his time out at the cabin. People wonder if he wasn't cheating somehow, ordering food in, or the nineteenth-century equivalent. People wonder especially today, when there is all this pressure on us to be cool to the environment, when there is a lot of pressure to be Thoreau.

Whether or not you buy his story, the Thoreau we know is a secular priest of solitude who lives quietly and alone and, frankly, prefers it that way. That Thoreau lives—and this is perhaps the most significant thing about the Thoreau we know—in nature, which is not like the place where the rest of us live. Thoreau's nature is a special, separate place, a place untouched by humans, except, of course, by Thoreau. The mere existence of the Thoreau we know stakes out the boundaries of this extraordinary place in our mind's eye, and in the eye of society. That Thoreau lives there makes him "inestimably priggish and tiresome," to quote a cultural critic.* In his own day, the *New York Times* described Thoreau as living in a "cold and selfish isolation from human cares and interests." James Russell Lowell, editor of the *Atlantic Monthly* when Thoreau was trying to get published, related him to Diogenes the Cynic, who lived on the streets of Athens in an open barrel, like a dog: "His shanty-life was a mere impossibility, so far as his own conception of it goes, as an entire independency of mankind. The tub of Diogenes had a sounder

* In this case the cultural critic is Bill Bryson, in *A Walk in the Woods: Rediscovering America on the Appalachian Trail,* an account of traversing the Appalachian Trail, or most of it.

bottom."* And yet as much as we chide him for being there, we don't necessarily want Thoreau to come home from the woods. Our relationship with him is symbiotic; the Thoreau we know, as difficult as he can be to deal with, helps us to live our daily not-so-strict lives, our less-intentioned existence. It's good to know the idea of him is still alive, that he is off in that special place, even if we don't spend much time in nature ourselves or haven't seen it since our last vacation, since we took a break from our personal everyday.

The Thoreau we know is a man of principle, steady, unbending, even when he applies himself to issues that, frankly, don't seem to have anything to do with nature or the environment, such as war. Indeed, he is the man who went to jail for his principles, who was carted off to prison for not paying taxes because he did not believe in an American military campaign, because he thought our war against Mexico was illegal, which it was. It might be argued, in fact, that we rely on his moral fiber—count on it, even—to live the lives we live. As citizens of everyday life, we maintain a measured and sensible flexibility, as opposed to Henry David Thoreau. We live in the real world, as opposed to a cabin off in the woods; in the real world, things happen, people have to be places, bills have to be paid, and eggs have to be broken. In a way, his intransigence allows for our flexibility. We can, in good or at least better conscience, bend in the winds of everyday practicality, precisely because of the trunk-like rigidity of Thoreau, the single-minded nature boy. He holds the environment

* The word *cynic*—as Thoreau, an indefatigable word scholar, would have known—is rooted in the Greek word *Kynikos*, meaning "like a dog," and was associated with his name via Norse mythology. "Perhaps I am descended from that Northman named "Thorer the Dog-footed," he wrote in his journal in 1852.

sacrosanct. Indeed, he is "the father of environmentalism," as Edward O. Wilson, the scientist and naturalist, has called him. He worships nature, monk-like, while we carry on at home, ministering to the demands of the nonnatural world. He tends the pure garden of Mother Earth, to be poetic about it, while we trudge through the fields of the mundane. There's even a hint of jealousy: while he gets to live in a cabin in the woods, we stay at home and go to work. We have to make a living. Though when our lives permit, we eventually turn and appreciate what he has done, who Thoreau is, and what we know he stands for (whether we have read a lot of his work or not). We behold the pure majesty of nature, we camp or hike or hotel in all its glory, then we get back into the car and go home.

BUT HERE'S ANOTHER THOREAU. Here's a Thoreau who lives in town, in the center city of Concord, which, while not quite the size of a city, even though it wants to be, is a large town. Here is the Thoreau who is born in town and, except for a few trips to the Maine and Massachusetts coast, except for a little less than a year in New York City, lives his entire life in his town. This Thoreau is a local yokel, playing in the woods as a kid, getting to know all the old farmers as he gets older. He goes to college in Cambridge, Massachusetts, a long walk, and then a few years later, a short train trip, away. While a student, he lives with a social reformer in a factory, one of the most radical reformers in America. He comes back to his hometown to discover that there are no jobs, a recession. He goes on the road, to Maine, and he can't find any jobs there either. Thoreau also returns to discover that Ralph Waldo Emerson—the most exciting intellectual and the most renowned intellectual reformer in America—is a neighbor, a neighbor whose sister lives at the Thoreau home. (To survive the recession, Thoreau's mother takes on boarders.)

Emerson and Thoreau hit it off, Transcendental mentor and pupil. They are friends in matters lofty (poetry, theology, critique of America) and in matters more practical—Emerson likes to go on walks in the woods, and Thoreau, having grown up there, knows all the places and even has a boat, until he sells it, in need of cash. Thoreau moves to Emerson's house, takes care of Emerson's children, his carpentry, his yard work, his gardening, all the while doing other chores for other people around the village, the Transcendental handyman. Thoreau tries poetry, then essay writing, then edits the Transcendentalists' magazine, the *Dial.* In none of these endeavors does he manage to make much in the way of money. He next tries his hand at a genre of writing that is becoming known as nature writing, and manages to achieve some commercial success, though Emerson is not as impressed with Thoreau's writing as Thoreau is. He moves to New York, tries to establish himself as a successful free-lance writer, but gets homesick and returns early to Concord.

When he comes home, he decides to build himself a little house on the pond on the edge of town, about forty New York City blocks from the village center on a woodlot owned by Emerson—a woodlot that is not so much woods, in the sense that we think of woods today, as it is a place where Emerson cuts the trees that each day heat his house as he writes away. When Thoreau's friends visit, when his neighbors and family come to the pond for picnics or to stop by for the watermelon party that Thoreau throws every year (which he inaugurated before his time at Walden and would continue after his time there), Thoreau is sometimes excited, sometimes in the middle of work, cranking out several essays and two books in two years. He put a chair out on the porch to let them know he was looking for company. The first book he writes at Walden, *A Week on the Concord and Merrimack Rivers,* is a failure, and it begins a rift in Thoreau and Emerson's mentor-mentee relationship—a rift that would affect the relationship between the two friends, and that would ultimately taint America's relationship with Thoreau and, I would argue, with its idea of nature.

> *My dear Henry,*
>
> *A frog was made to live in a swamp,*
> *but a man was not made to live in a swamp.*
>
> *Yours ever, R.*

Thoreau takes seven years to write and rewrite and rewrite his next book.

The next book is *Walden,* his best-known work in America and, along with his essay on civil disobedience, one of the most famous works of American literature in the world. Though there are a lot of people today who think of it as the work of a naturalist with a journal, it is something else entirely; it is a machine, a device intended to charge and change the reader, rather than incite a withdrawal from society. Yes, *Walden* does include observations on nature, lots of them. But Thoreau often worried about being a scientist, a collector of facts, a mere noter of the natural. With *Walden,* he is an artist, one who uses nature as his inspiration, the facts of natural life as his pigments. As a Transcendentalist, his art is didactic, and it seeks to inspire, to change; like the landscape painters of the Hudson River School, he sees the divine in nature, and presents it so that we may act accordingly.

To call Thoreau a nature writer is more than limiting, given the way that we tend to think about nature writing; Thoreau writes about the whole world, and he writes of Walden Pond so as to change the world, all worlds, natural and nonnatural alike, not that there ought to be any distinction. "We are acquainted with a mere pellicle of the globe we live on," he writes. "Many have not delved six feet beneath the surface, nor leaped many above it. We know not where we are." *Walden* is a work that intended to revive America, a communal work that is forever pigeonholed as a reclusive one. And what is perhaps most surprising is that it's a comedy; it's an economic satire draped in the language of nature and farming and the self-help books of the day that shows the mass of economic men to be a bunch of unwitting saps. With some disdain, Nathaniel Hawthorne referred to his Concord friend as "a humorist."

WALDEN DIDN'T SELL EITHER. It didn't do as badly as Thoreau's first book, but it was no huge hit. On at least one level, Thoreau didn't expect it to be: the second rejection is sometimes easier for a writer than the first. Thus, after *Walden,* Thoreau takes on writing as a kind of full-time avocation, working in his family's pencil factory, doing odd jobs while selling the occasional travel piece. He is a singer and a dancer. He plays the flute and likes to take his friends on moonlit walks and, despite his reputation, rarely seems to have gone on a camping trip alone. He is also a surveyor, helping house builders build, farmers settle their disputes. When people think of Thoreau, do they imagine all the time he spent in court, testifying to land boundaries?

And everything he learns, all that he considers, he puts into his spectacular literary achievement, his journal, the two-million-word collection of notes, thoughts, mental sketches—a mind thinking through the act of writing. He dies at forty-four. He may or may not have had books planned on wild apples, on the colors of leaves in the fall, on the day-to-day life of Concord, on the Indians of North America, whom he sympathized with in ways that were tinted by the racial bigotry of his time, but also in ways that recognized America's misguided tendency toward belief in a God-infused exceptionalism. He rejected Manifest Destiny and asked if the nation, like the woods, the ponds, the forests and fields of New England, might not be facing decline. He dies at home. His aunt asks him if he has made peace with God. He tells her he did not know that they had quarreled.

THOREAU IS CONFUSING. He was to his contemporaries. Today, adults force high school students to read him, though he critiques the life-in-a-rut grown-up and might prescribe a little teenagerness. He loved nature, but if we read him closely—or if we are down on Thoreau and trolling for critical ammunition—we see him cutting down trees, polluting ponds, working with land developers and miners, doing all the things that we associate with people who are not so Thoreau. Critics and commentators take the man who went to jail for his version of freedom and turn his critique into a prank, as proof only of his outsider status. "Civil Disobedience" is proffered as evidence of his unhealthy individuality, of his lack of humor and of compassion: as a civil disobedient he is an anarchist concerned merely with *numero uno.* Critics want to trap him, catch him out, and they do. Recently, a historian made this comment about a meeting of the Concord Women's Anti-Slavery Society held at Thoreau's cabin in 1837:

> This was an unusual incursion. Thoreau was an ardent abolitionist, *but one senses that he preferred jail to a cabin crowded with visitors* [my emphasis]. If *Walden* was Thoreau's flight from the market economy, it was, equally, a flight from women, from domesticity, from family life. He walked to town, nearly every day, to dine with friends; his mother often cooked for him. "I think that I love society as much as most," he wrote, "and am ready enough to fasten myself like a bloodsucker for the time to any full-blooded man that comes in my way." But he loved his solitude (a friend of his once said that he "imitates porcupines successfully"), and he hated hearing news. "Often, in the repose of my mid-day, there reaches my ears a confused tintinnabulum from without. It is the noise of my contemporaries." Above all, he cherished his manly self-sufficiency (even though he carried his dirty laundry to Concord for his mother to wash): "Who knows but if men constructed their dwellings with their own hands, and provided food for themselves and families simply and honestly enough, the poetic faculty would be universally developed?"*

A typical gotcha against Thoreau. But for my Thoreauvian money, it's misguided. Let's set aside the fact that his two years on the pond were just about the only time he did *not* live with his mother and his sisters, whom he was as

* This passage comes from a review in the *New Yorker* of *What Hath God Wrought: The Transformation of America, 1815–1848,* by Daniel Walker Howe. The review was written by Jill Lepore. I don't mean to pick on her. I could point to dozens of critics who say the same or similar things about Thoreau. Likewise, I could point to people writing from summer retreats, off in the woods, who extol the guy for his retreat to the woods. The point is critics as well as environmentally aware citizens living in the modern world are often trying to trip up Thoreau, to prove he cheated, or at least didn't do what he said he did.

likely to trade chores with as wildflowers and poetry. Let's ignore, for now, Thoreau's tireless collecting of local history at farms and taverns or when chatting up old-timers at the post office. Let's forget that he was, in fact, reading the newspaper accounts of explorers and politicians and politics just like everybody else in newspaper-rich antebellum America, that he was simultaneously savoring the papers and cursing them. (He had a subscription to the *New York Tribune.*) Let's forget that he was involved in and even dependent on the market economy that he supposedly rejected. Let's focus solely, in this particular gotcha analysis, on the assertion regarding his social proclivities. He preferred jail, this historian argues—or so it seems. Here Thoreau is criticized on two fronts: as a cipher, for preferring jail, and as a serial socializer, for walking to town with friends nearly every day. Gotcha! Contemporary critics are particularly peeved about his mother doing his laundry, at their boardinghouse—but it should be pointed out, first, that he didn't have much laundry (not that anyone else did), and second, that she seems to have had recent immigrants helping her with what had just been deemed housework—Irish laborers. Third, many people in America (including many critics) have someone else, possibly recent immigrants, doing their laundry for them, a phenomenon in the United States that kicked in around Thoreau's time.

And yet Thoreau is still somehow not quite commendable. He communes with nature, is a so-called tree hugger who, we feel, values plants and birds over people. And, even more confusingly, his comments are difficult to comprehend. He speaks mostly in funky parables that are difficult for us to recognize as old-style jokes: "And don't you ever shoot a bird when you want to study it?" a neighbor asked him once.

"Do you think that I should shoot you if I want to study you?" he fired back.

A lot of people, and not just high school students, think Thoreau is—to use a natural metaphor—a stick-in-the-mud.

It might be argued that he wrote too well for his own good, or for the good of his ideas, as his lines have been worked over so well, polished perfectly like the stones at the bottom of a stream, that they can be taken in without effort, even though they are meant to initiate work on the part of the reader. "In proportion as he simplifies his life, the laws of the universe will appear less complex, and solitude will not be solitude, nor poverty poverty, nor weakness weakness," he wrote in *Walden*. It's not difficult to find the aphorism that can be bent this way or that in the winds of rhetoric, but *Walden* itself can work like a giant aphorism—the reader runs the risk of perverting it into a cheesy scenic overlook on the road to what is profound. Like Mark Twain, who points out that we are going to hell as we laugh and laugh at him, Thoreau suffers for the smoothness of his wit, and what is radically paradoxical can be mistaken for advice on home décor: Simplify!

Thoreau is not a jerk *per se,* but he is jerky. He quits the church that his family attends. His family will eventually quit it too—it is a time when New Englanders are quitting churches, a time when people are quitting voting. Though it is rarely noted (and criticized even by Emerson), he will return for a sermon from time to time. More frequently reported is the way he once irritated congregation members by dragging a tree from the woods on a Sunday, passing the church's open front door, the congregants "gaping and horrified." He is a reserved and thoughtful critic, who goes quietly to jail to prove a point, but later he is a firebrand inciting violence, labeling John Brown, the armed revolution-inciting abolitionist, the exemplary just man, relating him to Christ, calling him the one Transcendentalist. When Thoreau sent a boy around Concord to tell people he was about to give a John Brown speech and word came back to Thoreau from the town principals that such a speech might not be appropriate, he told the principals that he had not been asking permission. Thoreau luxuriates in paradox. He likes a good fight, and often stands in loyal opposition. Here is the champion of civil disobedience, the inspiration of Mahatma Gandhi and Martin Luther King, Jr., extolling violence: "I do not wish to kill or to be killed, but I can foresee circumstances in which both these things would be by me unavoidable."

WITH SO MANY THOREAUS OUT THERE, it's difficult for us to know who Thoreau is for sure—radical or conservative, recluse or communitarian—but he is, without question, worth sorting through. He is worth sorting through, first, because of the depth of his art, his power as a stylist. He is America's proto–James Joyce, just in terms of wordplay. He gives birth to writers and thinkers as diverse as Edward Abbey and Martin Luther King, Jr. He gives birth to a lot of junk too. Think of all the blank journals sold at bookstores that go home with aspiring Thoreaus, think of all the solipsistic first-person novels that may or may not be transcendent. The irony is that the first person in Thoreau is the opposite of who it would seem to be. It is not an "I" at all. It is you.

What's confusing about the Thoreau we know, it seems to me, points to confusion about the way we live now. *What would Thoreau do?* is an oft-heard refrain. It's an understandable question, of course, but it's not worth pondering if we have a skewed perception of the guy. What he did is very different from what we generally think he did. The book in your hands may offer little new to the Thoreau experts among us (and there are many, as he is said to be the first American author to have a literary society established in his name). But I hope it will speak to people like myself who live with a conception of Thoreau that is a foundation to some of our fundamental conceptions about nature and the city, about life and civilized life.

This book is, thus, not a biography of Thoreau. It is a look at the times and the conditions under which he wrote, and a look at him as a free-lance writer, free-lance writing being one of the few areas in which I can be said to have some expertise. He was a professional free-lance writer, in fact, in the days when the notion of professional writing in America was just being established. I will not dwell so much on Thoreau's works themselves—I can't interpret *Walden* for you or anybody, especially since my own interpretations have developed as I wrote these pages. I'm just going to look at what is or can be confusing, keeping in mind always that Thoreau savored opposites, that the muddied swamp waters are as interesting to him as the pond waters that are crystal clear. Think of this book also as a snapshot of his time, which, as I look through its pages, bears a striking family resemblance to my own. In this way, I suppose, I am thinking not just about Thoreau, but about where America took him and why.

BECAUSE WHAT IF WE HAVE HIM ALL WRONG? What if the Thoreau you think of as a refuge-seeking mystic was a guy who, for a kind of a kick, built a little shack in the woods on the edge of town to save some money and get some work done on a book he'd been thinking about for a while, among other things? What if the new Thoreau is a humorist with the eye of a social satirist, who, when he read aloud from his works-in-progress on the trips he did take out of town, often had audiences in complete hysterics—as Emerson once described it in his journal, "They laughed till they cried." The Thoreau we know transforms from a crackpot to a trickster or a prankster, a member of, to be a little poetic about it, the cosmic opposition party. This is not to say that Thoreau was not a serious guy. He was serious. But after reading him for a while now, I think we may have missed a part of his humor, his whimsy, the stuff that makes him more human and maybe even more relevant to a lot of what's going on in our lives today.

The central question is this: If we think of Thoreau differently, will we think of his place—that is, nature—differently too? What are the implications of thinking that our first environmentalist was something else altogether, something much messier and a thousand times less monastic? How would we rethink the notion of civil disobedience if it turned out that Thoreau was not an anarchical loner? The way I see it, a drab seriousness drapes our thinking about the American Renaissance in general and Thoreau in particular, not to mention environmentalism. The seriousness manages to extend itself all the way into today, when we strip his words of their sometimes joyful intention, and end up with only the literal outline of his operation, a separateness that infects how we talk about the air and the sea, that makes us see these as places apart from us, as special.

It's my aim here to think about how, if the sacred was somehow mixed in with the profane in our view of Thoreau's ideas, our idea of what constitutes the environment might change. Among other things, I'd like to suggest that Thoreau was—aside from making the most serious points about the nature of life and the relationship between the individual and society and the individual and the world—having fun. I don't mean joking just to make people laugh; he was certainly not doing Transcendentalist stand-up. But he was joking to prod, punning to motivate, to inspire, to reform, in a world that he had a difficult time in but that nonetheless inspired him. "Surely joy," he said once, and over and over again in various permutations, "is the condition of life."

Not that we could ever actually nail him down—then or now—and not that we want to even try. Something to keep in mind about Thoreau is that he recognizes that there are stories outside the official story; he is the progenitor of alternate histories, a slayer (along with Emerson, believe it or not) of the story as written exclusively by and for the Dead White Male. But even shortly after he died, his friends saw that the Thoreau being preserved for posterity was significantly different from the Thoreau they knew. "Both admirer and censor have brought the eccentricities of Thoreau into undue prominence, and have placed too little stress on the vigor, the good sense, the clear perceptions, of the man," wrote his friend, Thomas Wentworth Higginson, in 1898. "I have myself walked, talked, and corresponded with him, and can testify that the impression given him by both these writers [his friend Ellery Channing and James Russell Lowell, the editor and critic who really, really did not like Thoreau] is far removed from that ordinarily made by Thoreau himself." This was a crime to those who knew him and sought to uphold his career and his reputation, but it also had other reverberations, as for instance on our perception of the natural world.

Thus, when we set out to meet the Thoreau we don't know, we are looking for him in another kind of place, a nature that is not necessarily far away, or even out of town, but a nature that is closer in, maybe a little less beautiful and certainly less simple. We might find him immersed in economics, writing about architecture, or discussing the emerging middle class. We might even find him walking the streets of New York with Walt Whitman.

I MET THIS OTHER THOREAU over a long course of time. I started out with the Thoreau everybody knows. I met him in my high school English class, reading parts of *Walden* and "Civil Disobedience." Growing up in a large metropolitan area, I imagined the landscape of *Walden* as being far away, in a place that I had never seen, and could only imagine—a wilderness. I grew up and went to college and lived in various cities, but visited places where the name Thoreau was repeatedly and reverently mentioned. I saw Maine, for instance, where his quotes decorate postcards of the mountains, and when I moved to the West Coast and hiked and camped in the relatively vast national parks and wilderness areas, I often heard his name invoked, even though he never visited there. On occasion, I myself have written books that were deemed nature books—I bristle at this particular categorization, though if you write books for a living you are happy to have them deemed anything—and I was sometimes called an urban Thoreau, a joke on somebody, I'm sure. The tag seemed like an oxymoron to me at first, but as I began to think about it, it didn't. I lived in a city and then for a short time in a suburb—and in the suburb I lived next to a little pond. I started reading Thoreau again. I read a lot more this time.

One day, as my interest in Thoreau was intensifying, I went to the New York Public Library, to the Berg Collection of rare books and manuscripts (and some other less bookish things), to see two Thoreau artifacts that in all my time in the New York Public Library, until recently, I had never known existed, never even imagined. I took the subway. It was a winter day, and I was bundled up so that on the subway the warmth caused me to become sleepy, and when we pulled into the Forty-second Street stop, I was just about sound asleep. Somehow—I never know how this works—I fumbled awake in the nick of time, and marched slowly upstairs, off to a quiet little room on the library's second floor. I presented my request slip, and in a second I was falling asleep again, yawning, struggling to keep my eyes open, while waiting for my items. After a while, a man brought out a map and a pencil, and my own pair of protective gloves. It was a little map of Walden Pond that Thoreau had sketched in pencil. The pencil was a pencil that Thoreau had made—that is, his family's pencil-making company had made. Thoreau had designed it. Thoreau took over the pencil-making factory when his father died, traveling to New York to make pencil-making deals, working out ways to make better pencils.

I held the pencil for a while. I didn't want to use it; since they aren't making them anymore, you have to conserve Thoreau's pencils. Also, that's probably against library rules. (It was fun to imagine what the sentence for writing a sentence might be.) But as I held it and looked at the lightly penciled map of Walden Pond, I wanted to shout with excitement, something that's surely also against library rules. I was feeling the exquisite happiness that comes from seeing something that I had never noticed before, despite it being right under my nose. I was seeing another Thoreau, out of the cabin by the pond and in a factory inspecting his pencils, walking the streets of New York and Boston, first with his father, then on his own, selling pencils—a thing so practical, so mundane! In their day, Thoreau's pencils were considered the best pencils in America.

I was in the middle of a city of eight million people, some running around crazy, in quiet desperation, some pausing for a moment, literally or mentally, and thinking about something—maybe even their place in the world, or in their own life. I was suddenly and completely awake, my idea of Thoreau rearranging itself, lines from *Walden* redefining themselves, as if under new ownership: "To be awake is to be alive. I have never yet met a man who was quite awake. How could I have looked him in the face?"

THAT'S WHAT THE OTHER THOREAU DOES TO YOU. When you begin to see him, he wakes you up. Hopefully, he will make you a little crazy too, and you will begin to see the woods in the town and the town in the woods, or at least you will begin to look for such things, even at home, in your own library. I have in my home, for instance, a beautiful coffee-table version of *Walden*. It is magnificently illustrated with dozens of beautiful, color-soaked photographs, each depicting Walden Pond and its environs as nearly pristine, almost human-free. On one page there is a lone fisherman in a halo of mist, and in an overhead shot we see the cut of the train track, overgrown and expired, and the faraway next town. In the distance, there is Hanscom Field, an airport, just about three miles away from Walden as the crow flies. It's the place where the Beatles landed in 1964, the airport used by President Gerald Ford when he flew in for the 200th anniversary of the battles of Lexington and Concord and the "shot heard round the world." Hanscom Field was all over television news in 2006 when reporters spotted the pitcher Daisuke Matsuzaka flying in to negotiate with John Henry.* I don't think the landscape, interspersed as it is with these new, seemingly un-Thoreauvian things, is any less Thoreau-related. I like to think the opposite: the human-touched landscape is a Thoreau landscape too. Meanwhile, the Thoreau you don't know is also a student of economics, parodying Ben Franklin's business advice: "Who would not be early to rise, and rise earlier and earlier every successive day of his life, till he became unspeakably healthy, wealthy, and wise?" as the "I" of *Walden* remarks on an early walk serenaded by birds in the woods—or what little, in Thoreau's time, remained of them.

I envision this Thoreau in Concord. He is living at home, in his parents' house in the center of town, where he pays them rent. A sister and a brother have died, but another sister remains and lives downstairs, while he lives in the attic, among his books—the entire Bhagavad-Gita, for instance, a gift from an English nobleman who had worked in vain to convince Thoreau to go to Europe—all stored on a shelf built from pieces of driftwood he'd collected at the river. There are his samples of leaves, shells, twigs, bones. There are the Native American archaeological finds and fossils, some of which he has carried home in the little compartment he made in the top of the hat he wears on long walks around town in the afternoon. There are his notebooks, dozens of them, with pages and pages of notes on the local flora and fauna, as well

* By John Henry, I don't mean the mythical figure of the working class, whose story represents the marginalization of workers in the changing nineteenth-century American economy, but the Red Sox owner and commodities trader whose story represents something else, something decidedly more capitalist.

as on the history of the town's buildings, the history of its people.

To make money, the Thoreaus take in boarders—they never know if they will stay afloat financially—and Thoreau's mother has the help of two girls, Irish immigrants like the other ones who are suddenly flooding into America. A visitor calls and Henry David is called for. Eventually, he comes racing down the stairs. He's dancing, a short-legged, slope-shouldered, lanky bachelor in his late thirties, with a big nose and soft gray eyes and rumpled hair. He is dancing a jig, his feet shuffling, kicking, a burst of joy, though he will eventually have to stop to breathe—heavy, tuberculosis-inflicted breaths.

"My Henry always was a good dancer," his mother will tell the visitors.

I see him dancing, singing, setting out to the factory with a skip in his step, a nod to the neighbor, a nod to me and you. This is the image of Thoreau I'd like you to start out with. The Thoreau you don't know is one of those rare writers who speaks in the prophetic voice that applies not so much to the future as to the place where the future is manufactured, where the past and present mingle, deep inside the mundane everyday.

Then again, I suppose I have an ax to grind. The Thoreau you know bothers me too, in light of the one I think I've seen. In fact, I could start with the ax. Because in order to build the cabin at Walden, the famously unsociable "I" borrows a friend's ax. He then proceeds to cut down some trees.

Chapter 2

WHERE HE WAS COMING FROM

FOR NEARLY HIS ENTIRE LIFE, Thoreau lived in town. His family moved away for a few years when he was a child, and he himself went away for four semesters of college in nearby Cambridge, taking a few weeks off to make tuition money by teaching at a minister's house thirty miles away from his hometown. He also lived for a little under a year in New York City, in Staten Island, to kick off his free-lance writing career. But other than those times away, he was always in Concord, where he was born in the summer of 1817.

Concord was (and remains) a proud town, the oldest white settlement in Massachusetts, a wilderness community in its time (and in Colonial thinking), situated at the confluence of the Assabet and Sudbury rivers, where the Concord River begins. In Thoreau's day, it was surrounded by farm fields in gentle rolling hills, low mountains in the distance, the ocean a day's boat ride away. Once, it was the site of a Native American settlement, Musketaquid, thought to have been wiped out primarily by smallpox. Concord was named for what the English settlers liked to think of as a peaceful transaction, the purchase of the land from the remnant population of native people. (They were likely remnants of the survivors of King Philip's War, a not-at-all-peaceful battle that pitted local English settlers against Indian tribes in the area, the latter led by a Native American leader known to the colonists as King Philip.) Concord is and was a capital of the myth of America. It was famous in antebellum America, as it is now, as the site of what was the first military engagement of the American Revolution, as the birthplace of the Minutemen, the volunteer militia of farmers, as the setting of the aforementioned "shot heard round the world" and the midnight ride of Paul Revere. Originally, the battle was remembered as a bloody slaughter of ragtag insurgent farmers, fighting to conserve their rights as British colonists against a well-financed military—even George Washington was critical of the volunteer militias—but time and repeated tellings changed it to a more noble face-off between America's honorable, freedom-inspired homegrown military and a vast reactionary power. Concordians thought of themselves as being at the birthplace of American independence, as did Thoreau, and eventually he would challenge them to think about being revolutionary one more time.

In 1837, the year Thoreau graduated from Harvard, Concord dedicated its Revolutionary War monument, after much debate and several postponements. Concord had not always celebrated its battlefield, and from time to time, the statehouse in Boston quashed its funding in order to subdue Concord's occasional attempts to become the state capital. It was a contentious anniversary in Concord, because of lingering resentments stemming in part from the time the Marquis de Lafayette had visited in 1824 and angry residents were held back from a small private dinner tent on the public square by the bayonets of soldiers, while the marquis and a select Concord few celebrated inside. Ralph Waldo Emerson, a controversial Unitarian minister who had recently resigned from his church and moved to town to write and lecture, was invited to celebrate the 1837 dedication—Emerson's family stretched back to the founding of the town. Emerson wrote "Concord Hymn," an elegy to the battle and the town which helped make the gunshot fabled. It was sung to the tune of "Old Hundred." Emerson himself could not personally attend the dedication, but Thoreau sang the song as a member of the choir:

By the rude bridge that arched the flood,
Their flag to April's breeze unfurled,
Here once the embattled farmers stood
And fired the shot heard round the world.

When Thoreau had been about to set off for college, Concord had been a lively place. There were two banks, forty shops, and four hotels. It was a four-hour stagecoach ride to Boston, and a short time away from becoming a one-hour train ride; over the course of just a few years the railroad was to change American life, the way the interstate highway system would in the twentieth century, and today, as far as transportation goes, Concord seems like a lot of places in America—a little old downtown that's just off a state road within striking distance of one or more interstates. But before the railroad arrived, Concord had been a transportation hub, at the intersection of old and well-used roads that connected it to Boston to the east, New Hampshire to the north, and, to the west, the Berkshires and the rest of western Massachusetts. The taverns were often filled with teamsters, the horse-equipped truckers of their day, who carried goods into the burgeoning market outside of Concord, a relatively new development on the economic landscape.

In 1837, with a population of two thousand, Concord was a medium-sized town that hoped to be bigger but at the same time still liked to think of itself as an amiable, civilized village, even though the strains of the new economic order were already apparent. Wage earners began to be distinguished from non–wage earners, for instance, which among other things began to affect how people lived together, as well as what neighbors thought of as neighborly. Rev. Ezra Ripley was the senior pastor of the First Church in Concord; he baptized Thoreau, and though he sometimes seems to be at odds with everything Thoreau will eventually stand for, he can also seem to be writing and preaching on the same themes as his one-time parishioner, as he does here, commenting on the cracks in Concord's civic life: "With more than a few, it has been too much the practice of neighbors and fellow citizens to live like strangers, and to cherish little or no sympathy from one another. One class of citizens holds themselves at a distance from another class,—one individual from another." By the 1890s, after Thoreau had died, Concord's economy would be completely transformed: the factories closed, farmers were gone, the stagecoach lines no longer stopped there, and people commuted to Boston on trains, to work *away*. A way to think about what was happening then is to imagine an old downtown now, at the time that another big superstore has just moved in out by the interstate highway, near all the other superstores, as the commuters are beginning to not shop near home anymore, to no longer associate. "It is a dull place," a resident would say of Concord a few decades after the Civil War. "It is a narrow old place. It is a set old place. It is a snobbish old place. . . . It is full of graveyards, and winters are endless. The women never go out, and the streets are full of stagnation."

At Thoreau's birth, Concord, like many towns in New England, had been in the midst of a shift from self-sufficient farming village to manufacturing town; in the village, little shops made shoes, lead pipes, clocks, hats, bellows, guns, bricks, barrels, soap, and pencils, almost all sold outside of Concord, shipped out on the roads and river. There was still a textile mill. It was a time of tremendous economic flux, a situation Thoreau will address when he writes *Walden*. Sons could not count on doing the same type of work their fathers had—in fact, they could not necessarily count on work. "Modern" farmers hired out work that might formerly have been done by family members or neighbors. The distance between those who had wealth and those who did not was expanding; sixty-four percent of the town did not own land, and about fifty men controlled roughly half the wealth. Like most of the United States at the time, Concord was about to suffer a major economic downturn, just as Thoreau and his contemporaries were about to graduate from college.

In 1837, Concord was also becoming known as the headquarters of an intellectual movement, small in number but intense in effect, called the Transcendentalists. "Why, you're the biggest little place in America—with only New York and Boston and Chicago, by what I make out, to surpass you," Henry James would say, years later. It was also becoming a suburb, as the idea of a suburb was coming into existence. It was, perhaps, a suburb first for writers, who desired access to Boston, twenty miles away, but also wanted to enjoy walks in the rustic environs, to revel in the notions of rural romanticism that had spread from Europe to the United States. Concord was also what might be called a radical suburb, an idea that was not yet an oxymoron. Concord was an idea factory, and the factory's foreman was the local celebrity, America's first public intellectual, Ralph Waldo Emerson. Thoreau was about to become his apprentice.

To Thoreau, who would toy with leaving but in the end choose to stay, Concord was a good example of what he would later refer to as "a partially cultivated country." In 1853, in a piece on Maine, he wrote, "The wilderness is simple, almost to barrenness. The partially cultivated country it is which chiefly has inspired, and will continue to inspire, the strains of poets, such as compose the mass of any literature." He enjoyed visiting wilderness, but the partially cultivated country was always his preference. Thoreau was, as the historian Robert A. Gross has written, "a product of the central village."

THOREAU'S FAMILY'S ROOTS WERE in New England and France. On his father's side were Huguenots who had left France after the Edict of Nantes, settling in the Isle of Jersey. His paternal grandfather became a privateer, a nice way of saying *pirate* in the eighteenth century; he traded and made a small fortune, so that when he died, in 1801, he left a $25,000 estate, which was slowly mismanaged away. On Thoreau's mother's side were Loyalist elites, the Dunbars, who showed the same kind of civic spunk that Thoreau and his mother would be chided for. While living in Concord during the Revolutionary War, four Dunbar brothers were arrested and jailed for their support of the king, and their estate was seized. Thoreau's maternal grandmother, Mary Jones Dunbar, smuggled in tools with the food; using files to remove the bars, the four brothers broke out of jail and fled to Canada. When her first husband, a minister, died, Mary married a landed Concordian and moved into the farmhouse in which Thoreau was born. By the time Thoreau came along, he had a sister, Helen, and a brother, John, and the family money was mostly gone. The family often lived in what Walter Harding, Thoreau's preeminent biographer, has described as poverty.

Until Thoreau was five, his family was itinerant. Thoreau's father, John Thoreau, attempted to run a grocery store in Concord, which failed, and then a grocery and liquor store in Chelmsford, Massachusetts, ten miles away. Very soon after Thoreau was born, the Chelmsford store failed as well. John Thoreau is said to have been too liberal with credit, and too lax in collecting it—not a bad guy, in other words, though soft by the business standards of the time. The way that people were running businesses was changing; it was around this time, for example, that grocers began charging particular prices on items. (Previously the idea of pricing was more vague.) Some biographers say Thoreau's father sold his wedding ring to pay the creditors. But John Thoreau was likely a victim of the times, which were bad for everybody. The Thoreaus' last move before returning to Concord was to Boston, where John taught school. While there, they continued to visit Concord, to see Thoreau's mother's family. Thus, as it would one day be for America, Walden Pond was already a romantic ideal for the young Thoreau, a place where they went for picnics, fishing, to make chowder. As an adult, Thoreau reminisced in his journal about visiting Walden Pond for the first time when he was five. "That woodland vision for a long time made the drapery of my dreams," he wrote.

The Thoreau family's financial luck changed for the better when Thoreau's maternal uncle discovered a deposit of graphite (then called "plumbago"), the stuff of pencils. He staked a claim, and decided to become a pencil maker. John Thoreau moved back to Concord to work with him in the pencil-manufacturing business. Pencil making ended up being a business that Thoreau's father was pretty good at. This time the Thoreaus managed to stay in Concord, finding a niche in the new commercial environment and, in so doing, becoming part of a relatively new thing in America: the striving middle class.

Though successful at last, Thoreau's father is described in most accounts as a relatively low-key pencil maker. He loved to fish and walk at the pond, to make chowder with his kids. He loved music and played the flute in church. He passed on his love of music to his children, and he passed on his wooden flute, which ended up with Thoreau (and is now in the Concord Museum). Once, those pursuits had mostly been valued; by 1859, they were less so. It was as if a huge portion of his personal portfolio gradually evaporated. Thus, the elder Thoreau was eventually described as "an amiable and most loving gentleman, but far too honest and scarcely sufficiently energetic for this exacting yet not over scrupulous world of ours." In other words, he was not considered—to use a term that was ubiquitous in early industrial America—sufficiently "enterprising" for the burgeoning commercial society.

When John, Sr., passed away, three years before Thoreau

himself did, the son reminisced fondly about the father in his journal, describing the old man's quiet death and long life in Concord, as well as his stints in Boston and Maine. At meals throughout his entire life, Thoreau had listened to his father talk about having breakfast with *his* father, sharing a roll. Thoreau heard about his grandfather losing his job as a cooper at the commencement of the Revolutionary War and going to war as a privateer, which we might think of today as a free-lance military contractor—that is, he fought on the American side in part because that was where the money was. Thoreau's grandfather had served under Paul Revere, who was in charge of defending Boston Harbor; he was a link to another era, another America. As his short life progressed, Thoreau valued his own family's and his town's history and, in a sense, his parents, more and more; he valued those links back to a revolutionary era. He became more and more fascinated by the history of Concord, natural and otherwise; he began to see Concord's history, in fact, as less like a time line and more like a layer cake. His father, he knew, could perceive many of the very different layers, a skill Thoreau valued more and more as he got older, because when you get older, you tend to value more about your father.

"He belonged in a peculiar sense to the village street; loved to sit in the shops or at the post-office and read the daily papers," Thoreau wrote in his journal at the time of his father's death. "I think that he remembered more about the worthies (and unworthies) of Concord village forty years ago, both from dealing as a trader and from familiar intercourse with them, than anyone else. Our other neighbors, now living or very recently dead, have either come to the town more recently than he, or have lived more aloof from the mass of their inhabitants.

"As far as I know, Father when he died was not only one of the oldest men in the middle of Concord," Thoreau continued, "but the one perhaps best acquainted with the inhabitants, and the local, social, and street history of the middle of the town, for the last fifty years."

THE THOREAUS WERE NOT ALOOF. People in town who didn't like the Thoreaus might have thought they were aloof or called them aloof, and that point of view still thrives today, in a genetic chain of irritation—bitterness has a longer half-life than kindness and common courtesy, it seems.* But there were a lot of people who wrote letters and noted in their personal journals that Thoreau and the Thoreaus were not so bad, or were good even. Indeed, it is not too difficult to draw a picture of the Thoreaus as the opposite of aloof—as connected, secured to Concord and its doings, or at the very least "acquainted with the inhabitants." The Thoreaus were what we might today call "active in their community,"

* A central theme that anyone considering Thoreau must face early on is the jerk factor. Was Thoreau a jerk? Recently, I met someone in New York City who had a very good friend in Concord who knew for a fact that Thoreau was a jerk. "Everybody hated him," the guy told me. This is not the case; *everyone* did not hate him. The Concord resident who brought him a skeleton found inside a wall in 1853, for instance (a visit Thoreau notes in his journal), does not appear to have hated Thoreau, or if he did, he didn't hate him enough not to stop by: "Mr. Pratt asked to what animal a spine and broken skull found in the wall of James Adams' shop belonged." Thoreau thought muskrat. Pratt disagreed. Thoreau used his microscope to examine the teeth. "I told Pratt it was a muskrat," Thoreau wrote, "and gave him my proofs; but he could not distinguish the three molars even with a glass, . . . for he had thought them one tooth, when, taking out his pincers, he pulled one out and was convinced, much to his and to my satisfaction and our confidence in science!" If a jerk, Thoreau could be described as a helpful jerk at the very least. A neighborly jerk even, and who among us does not have a jerk for a neighbor? Or vice versa. Put another way: *Who are you calling a jerk?*

a phrase that still sounds like a worthwhile goal to us, despite the current tendency of many Americans to be less than active, due to time and money constraints. In addition to her work as an abolitionist, Cynthia Thoreau, Thoreau's mother, was vice president of the Concord Female Charitable Society. ("The chattables," as Thoreau liked to refer to them.) Cynthia held various social gatherings at their home, which eventually, as they gradually earned more money from pencil making, was situated in the center village. In the various houses that they rented and lived in, the Thoreaus held so-called teachings, book-group-like gatherings formed to discuss abolition or other causes. On June 17, 1853, Thoreau wrote in his journal: "Here have been three ultra-reformers, lecturers on Slavery, Temperance, the Church, etc., in and about our house and Mrs. Brook's the last three or four days." The Thoreaus celebrated Christmas; in an era when Christmas was still escaping the sanctions of Puritanism, Santa Claus filled the Thoreau kids' stockings. Judging from letters and the journals of others, they threw a lot of parties. They had their piano; there was singing and dancing. People in town remarked favorably on the house parties thrown by the Thoreaus.

Music was a bigger part of everyday life then than it is now—homemade music, music at dances, music sung at work and in the fields—and all through his life, Thoreau danced, sang, and played the flute, writing in his journal frequently on this matter:

> *Music is the sound of the circulation in nature's veins. It is the flux which melts nature. Men dance to it, glasses ring and vibrate, and the fields seem to undulate. The healthy ear always hears it, nearer or more remote.*

> *When I hear music, I flutter, and am the sense of life, as a fleet of merchantmen when the wind rises.*

Nonetheless, the popular conception of Thoreau is devoid of music, except for the drum, thanks in large part to the famous line from *Walden* that has become a staple at graduation ceremonies: "If a man does not keep pace with his companions, perhaps it is because he hears a different drummer. Let him step to the music which he hears, however measured or far away." (The other graduation staple: "Go confidently in the direction of your dreams. Live the life you have imagined.") Thoreau's drumming metaphor is readily misinterpreted. First of all, before he went to Walden Pond, Thoreau marched with the local Concord militia. Second, it bears noting that marching to any drummer at all is not a solo act, unless you are the drummer, which Thoreau was not.*

His mother is said to have scrimped to buy the family a piano, just as she also seems to have encouraged his interest in the flora and fauna of the town—botanizing was all the rage

* Garrison Keillor, who has written glowingly of the writer E. B. White, who had written glowingly of Thoreau, wrote un-glowingly of Thoreau in his syndicated newspaper column:

> The philosopher of cheerful purpose was Emerson, and for some reason my generation preferred the puritanical Thoreau, a sorehead and loner whose clunky line about marching to your own drummer has found its way into a million graduation speeches. Thoreau tried to make a virtue out of lack of rhythm. He said that the mass of men lead lives of quiet desperation. Okay, but how did he know? He didn't talk to that many people. He wrote elegantly about independence and forgot to thank his mom for doing his laundry.

E. B. White spent some time writing in a shack, and while he was shy and took time away from the big city to write, you never get the feeling that he is a sorehead and a loner. Neither, I would argue, was Thoreau.

at the time. Thoreau could be a grumpy son; imagine living with someone who had to get out to check the river levels every day, who brought in plant specimens day in and day out, whose neighbors thought of him whenever they had a dead animal they wanted to figure out. But he was very often appreciative of his mother's expertise; in his journal, he notes that when he was searching for a particular plant in Concord, his mother was the one who told him what hill to find it on. Sometimes they wrote letters to each other in Latin, for fun. Thoreau's sense of civic duty—and, yes, it's true, he had one—seems to have come in part from his mother, who was, according to Harding, a person neighbors could turn to when they were low on food, which was often the case in the days of Henry Thoreau. The Thoreaus are reported to have aided Native Americans and the numerous Irish immigrants who were in need, and they took in slaves on the run to freedom in Canada at the end of the Underground Railroad.

Thoreau's mother also seems to have cultivated her son's awareness of sounds in general, an excellent habit of active listening that is still rare today. "My mother was telling to-night," Thoreau wrote in his journal, on a spring day in 1857, "of the sounds she used to hear summer nights when she was young and lived on Virginia Road,—the lowing of cows, or cackling of geese, or the beating of a drum as far off as Hildreth's, but above all Joe Merriam whistling to his team, for he was an admirable whistler. Says she used to get up at midnight and go and sit on the door-step when all in the house were asleep, and she could hear nothing in the world but the ticking of the clock in the house behind her." (One way to think about Thoreau is to imagine him sitting on the doorstep of all America—economic and natural—and describing the sounds of the modern world.)

Just as Thoreau's father is often cast as a quiet simpleton, so Thoreau's mother takes a lot of grief in the Thoreau historiography as shrill and harsh, a criticism of outspoken women even now. Thoreau's younger sister, Sophia, may have inherited her mother's tendency toward frankness. She is said to have marched out of her church, for instance, rather than take communion when she did not believe in it. Sophia would often go boating with Thoreau, on the river and Walden Pond, and as he would paddle, she would sketch. She played piano and painted, and gave lessons in both. Later, Sophia would edit her brother's papers, and when he was sick and moved downstairs from his room in the attic to the Thoreaus' parlor, she helped him write his last letters. She was a founder, along with her mother, of the Concord Women's Anti-Slavery Society, which met at least once at Thoreau's cabin at Walden Pond during the short, two-year period that he was living there. She helped run the family pencil business after Thoreau died.

Thoreau's sister Helen was also a teacher, in Roxbury. She had moved away from home to find work, and Roxbury was growing in population as an industrial town. Like Thoreau, she was plagued with tuberculosis, which the Thoreaus seemed to have initially contracted in their late teens. For a short time, she and Thoreau toyed with the idea of teaching at a school together, in New York City, until her ill health forced her to move from Roxbury and live at home in Concord. She then taught at home, while Sophia gave piano lessons. She died in 1849 at the age of 36. Her funeral was at the Thoreau home, with local Unitarian and Trinitarian ministers in attendance. It is said Thoreau sat quietly through the service, and just as the pallbearers were about to take away the coffin, he wound a music box and let it play out its tune. An obituary extolled Helen's work as an abolitionist and described her as "quick to feel the woe of others," adding: "She had the patience to investigate truth, the candor to acknowledge it when sufficient evidence was presented to her mind, and the moral courage to act in conformity with her convictions, however unpopular these convictions might be to the community around her." Thoreau wrote a poem:

Whether by land or sea
I wander to and fro,
Oft as I think of thee
The heavens hang more low . . .

Both Sophia and Helen were well known among the New England abolitionists. It is likely that people knew Henry David Thoreau as the brother of Sophia and Helen, and as the son of Cynthia. Competition breeds in families like weeds in a field, but traces of competition in the Thoreau family are difficult to find. Thoreau seems to have gotten a lot out of being a Thoreau. In 1840, he made a pun on the *family* in *familiarity,* writing to Helen in a letter: "This much, at least, our kindred temperment of mind and body—and long 'family-airity—have done for us, that we already find ourselves standing on a solid and natural footing with respect to one another."

UNTIL THOREAU WAS TWENTY-FIVE YEARS OLD, his best friend was his brother, John, Jr. John was the star of the Thoreau family. He was considered by the family and by people in Concord to be smart and collegial—"his father turned inside out," according to a neighbor. Where Henry could be quiet and reticent, John was gregarious and charming. John was the class clown, easygoing, the Thoreau boy whom everyone in town expected great things from. Henry loved spending time with his big brother—hunting for arrowheads, camping, playing on the rivers and in the ponds that decorated the area. They were best friends, true brothers. It seems safe to say that Thoreau admired his brother as much as did the rest of his family and community, and they worked well together, thriving in their differences. Henry was cast as a budding iconoclast, quiet and retiring in comparison. After Thoreau died, it was recalled that even as a child he had been a stoic. As a schoolboy, his nickname was "Judge."

The stories that survive in small towns and families are like the stories of the lives of the saints: the farther back they go, the more meaningful they become, and lives become almost pure in retrospect. "Mother, I have been looking through the stars to see if I couldn't see God behind them," Thoreau is said to have said to his mother as a child. It is said he did not want to go to heaven because he feared he could not take his sled there. It is said that he once asked his elementary school teacher, "Who owns all the land?"

Chapter 3

READING TRANSCENDENTAL

THE 1830S WERE ONE of those times when America needed a kick in the pants and a lot of people knew it, though all those people had very different ideas of where and how to deliver the kick, resulting in no one effective boot. Party politics had gotten to be a too-big party, such that political participation seemed mostly to be about mass spectacle; political machines seemed to be without anyone at the controls. The United States wasn't a young country anymore; it was a country that was middle-aged and facing market changes and societal changes that were causing people to fear they were losing control of their lives, or at least their lives as they had known them for a few generations.

Thoreau enrolled at Harvard in 1833, the year Andrew

Jackson was sworn in for his second term, the year the English Parliament abolished slavery while the American government yet again passed on the question. Eighteen thirty-three was a bad year for Harvard, even. Small and parochial, Harvard's courses and educational ethic had been unchanged since prior to the Revolutionary War. During Thoreau's time, there were student riots—over food and housing conditions—marked by smashed windows and furniture. As far as scholars can determine, the riots did not include Thoreau, though the president noted that Thoreau had "imbibed some notions" that the other students had imbibed. The classes were criticized too; the classics were taught by rote, with very little along the lines of elucidation. Anyone who has dealt with the test-based American school system in the twenty-first century or watched a child undergo standardized testing, an alleged indicator of higher understanding, would recognize Thoreau's criticisms. Emerson, who loved to egg on his young trainee's sarcasm, once argued to Thoreau that Harvard taught all the branches of knowledge, to which Thoreau replied, "Yes, indeed, all the branches and none of the roots." In fact, Thoreau "barely got in," according to a letter from the president; though when he did get in, he maintained above-average grades.

There are the standard two Thoreaus at college: the Thoreau who is perceived as friendly and the Thoreau who is considered cold. John Weiss, a classmate of Thoreau's who wrote poetry for school publications and later became a minister and author, wrote: "He passed for nothing, it is suspected, with most of us; for he was cold and unimpressible. The touch of his hand was moist and indifferent, as if he had taken up something when he saw your hand coming, and caught your grasp upon it." (Weiss, a writer, wrote this *after* Thoreau had achieved some notoriety as a writer, specifically as a social critic.)

Thoreau's friends, on the other hand, liked Thoreau, writing him letters when he was away trying to raise money for tuition or when he was home sick—his family wondered if his health would even allow him to finish school, since the tuberculosis had begun to take a toll. On one occasion, he left school to take a trip to New York City with his father to sell the family pencils, enough to make tuition expenses. He returned, writing a mock formal letter to a friend, asking to be his roommate: "Your humble servant will endeavor to enter the Senior Class of Harvard University next term, and if you intend on taking a room in College, and should it be consistent with your pleasure, will joyfully sign himself your lawful and proper 'Chum.' "

In school, Thoreau read the *Georgics,* the Roman poet Virgil's version of a farm journal, the simultaneously bright and brooding epic that is about agriculture—four books, covering crops and cattle and beekeeping. It is an intensive study of what today we call *place* and our relationship to it but that in Thoreau's time we would have called the *land* and *toil*. It is a work that is *Walden*-like in its localness, as well as being a work that is ostensibly about one thing (farm life) but actually about another, which is something Latin scholars debate: perhaps the vulnerability of man in a fallen time, perhaps man's attempt to understand his place in the physical world—how our own nature aligns with the nature of the world. Thoreau fell in love with it.

Reading the *Georgics,* Thoreau was inspired by the task of writing, the craft of it, as it related to the rhythm of agricultural life; he became enamored with the idea of the seasons as an artistic conceit, a metaphor that he would perpetually mine, that would protect him from his own depression, maybe even from the tuberculosis that rarely dragged him down. He was forever the poet farmer, minus an actual farm, which was emotionally reassuring for him.* Thoreau began to see the freshness of spring, for instance, as unvarying over generations—the very charge of spring, the vernal engine. He finds in the *Georgics* what he considers a noble zest for the daily occupations—life as tasks that can be attacked with vigor, if not joy. He read the Roman poet and satirist Horace's *Odes*. "Whether the gods allow us fifty winters more or drop us at this one now which flings the high Tyrrhenian waves on the stone piers, decant your wine," Horace wrote.

Ironically, at about the time he was deciding how to live, the disease that would eventually defeat him had entered his body. Tuberculosis meant coughing, of course, bloody sputum, loss of weight, and fatigue. It would mean that he would have trouble breathing whenever he had outbreaks over the years. It was a disease that regularly depleted New England villages. Tuberculosis would have caused him to have trouble sleeping and to lose his appetite for long stretches. And tuberculosis doesn't strike just the lungs; Thoreau may have had pain in all his joints, his muscles, anywhere—it took a lot of strength to be as sick as Thoreau was and still write as much as Thoreau did for so long. What's certain is that it slowly spread through his body, sometimes crippling him, starting when he went away to college.

Though his health had barely allowed him to graduate, he felt the need to decant. With the ancients covering his

* "I'm not fatalistic," Bob Dylan said in a 1965 *Playboy* interview. "Bank tellers are fatalistic; clerks are fatalistic. I'm a farmer. Who ever heard of a fatalistic farmer?"

back, he was ready to march into the promising future. He returned to Concord, where, since he had left, a railroad had come in, young men and women had moved away, farmers had been giving up their farms. "Three thousand years and the world so little changed!" Thoreau remarked to his journal.

As dead as the curriculum at school seemed to Thoreau during his time at Harvard, the air in Cambridge was alive with intellectual ferment. If you were a young man who fancied himself a poet, as Thoreau did, if you were infused with the poetry of the ancients and charged up by the modern experimental writings of the Germans and the English Romantic poets, and if you were hoping that American writers might be able to get something new going, something particularly American, then you couldn't pick a better place to be in 1837 than at Harvard, where a new intellectual movement was about to be born. It's like wanting to be a Beat poet and having Allen Ginsberg as your commencement speaker. Thoreau graduated at nearly the same moment that an intellectual movement was born, the Transcendentalists—a small and eclectic group of ministers, writers, and intellectuals who can today seem a discountable group, designed solely as the answer on a multiple-choice English exam.

The Transcendentalists' most famous proponent was Ralph Waldo Emerson, the Transcendentalist Philosopher King, who is often remembered as the writer who encouraged America to pull itself up by its bootstraps, as creating the culture of the self-made "Capitalist Man." This is not quite right. In fact, the Transcendentalists in general were thinking critically about society, and, despite their brief tenure—they were barely a group during the twenty years between 1830 and 1850—they are one of the most important intellectual groups in the history of America, having influenced the culture in the manner of the American expatriates of the 1920s and the Beats of the 1950s, and yet they can seem irrelevant today, inducing not a spirit of reform but sleep.

The Transcendentalists suffer now, to some extent, from the same public relations problem they suffered from in the 1840s—when they were a movement that was trying not to be a movement, when someone else called them Transcendentalists and they reluctantly agreed. (They sometimes referred to themselves as "the club of the like-minded," despite the fact that they rarely saw eye to eye.) Their detractors saw them as loafers, effete intellectuals who did not contribute to society in any gainful way and were at best critical, at worst removed. Even their intellectual relatives ridiculed their social experiments. Thomas Carlyle, the English writer who had inspired the Transcendentalists with his own writing, described George Ripley, the Unitarian minister who started the utopian farm community Brook Farm, as a minister who had "left the pulpit to reform the world by cultivating onions."

They were obvious targets for lampooning in that they were breaking conventional religious strictures, with their appeals to find God in nature, or, at the very least, to find God outside the traditional settings of the church. A cartoon printed at the time pictured a Transcendentalist as a human being who was mostly head, and the head being mostly one large eyeball, Emerson's large, all-seeing *I*. It showed the Transcendentalist standing outside in the rain, the caption: "I am raining."

The Transcendentalists are problematic, to say the least: they attacked Puritanism with a puritanical zeal, and they could never say for sure if the soul could handle all the work they planned for it. What's often overlooked when we look back at them today, however, is their role as social reformers, as activists who, as literary historian Philip Gura writes, "cultivated a vibrant openness to social and cultural ideals that directly challenged the materialism and insularity that were already hallmarks of American culture."

THOREAU WANTED TO WRITE—he knew it by the time he had graduated—and while it is possible to be an antisocial writer, a writer who is uncomfortable in crowds, who trembles at the lectern, who prefers seclusion to everyday human interaction (not that Thoreau fell into any of these categories), most writers have a desire to be published, and publishing is a social act. As Thoreau sets out to write and ultimately to be published, he enters the publishing world as a Transcendentalist, as a disciple and, at first, an imitator of Emerson. The Transcendentalists who came to Concord to study and walk in the woods with Emerson were Thoreau's philosophical comrades and his rivals, thinkers he agreed with in principle, but argued with in practice. They all argued, in fact. Emerson, especially, liked them to challenge one another, to criticize and bolster. To be a Transcendentalist was to egg the others on to a more considered position, to transcend, and there were a lot of different positions within the Transcendentalist tent.

Margaret Fuller, who rowed with Thoreau on Walden Pond, was an American protofeminist. "[N]ot one man in the million, shall I say," she wrote in the *Dial,* in 1843, in an article entitled "The Great Lawsuit: Man versus Man, Women versus Women," "no, not in the hundred million, can rise above the view that woman was made for man." In her book, *Women in the Nineteenth Century,* she envisioned the United States as the one country where women might be equal with men. Later, as the first full-time female columnist for an American newspaper, the *New York Tribune,* she did not cover home decorating: she was the literary critic, as well as a social critic, a muckraker.

There was Theodore Parker, the abolitionist Unitarian minister, who inspired Abraham Lincoln and later Martin Luther King with what was called the Social Gospel movement, which saw the remedying of social injustice as the practice of one's Christian faith. In "A Sermon for Merchants," given in 1846, Parker chided a feudal theocracy and "a beggar sect" and called for a government "of all, for all, by all," a phrase Lincoln used in the the Gettysburg Address seventeen years later. "Aggressive war is a sin, a corruption of the public morals," he said, in response to America's attack on Mexico.

The Transcendentalists worked together in their various experiments in communal living, associations aimed at offering an alternative to what they saw as industrial capitalism's threat to community, and yet they specialized. Elizabeth Peabody, an educational reformer, was, late in her life, a lonely voice against the relocation of Native Americans to the land of the Louisiana Purchase. (Thoreau's sister Sophia was also ardently opposed to Indian relocation and is thought to have helped convince Emerson to publicly oppose the plan.) George Ripley, prior to starting Brook Farm, was a minister in Boston, living and preaching in a working-class neighborhood. He became more interested in issues that touched the lives of his parishioners, issues regarding jobs and income and work that his church seemed to be retreating from. He feared "the inordinate pursuit, the extravagant worship of wealth," which, as he saw it, was driving the country. Bronson Alcott, also an educational reformer, considered children to be "a Type of the Divinity," and sought to make teaching less rote and more of an educational experience, for the teacher *and* the pupil. He lived in Concord and was a close friend to Emerson and Thoreau. Emerson considered him a great conversationalist whose ideas would not convert to print. Each time Emerson would rewrite one of Alcott's long and barely intelligible essays, Alcott would rewrite it to make it barely intelligible again—when Emerson writes about Alcott in his journal, you can almost see the Sage of Concord rolling his eyes. Before Thoreau did, Alcott went to jail in Concord for not paying his taxes, and even Thoreau, considered by many in Concord to be eccentric, considered Alcott to be eccentric.

And then there was Emerson, Thoreau's mentor, whom I will get to shortly, but who is perhaps best known, as I mentioned, for his insistence on an individual's self-reliance, by which he means, among other things, a reliance on the activation of one's own soul, an interior awareness of the world's divinity and the divinity inside one—i.e., you. To understand what Thoreau was thinking when he wrote *Walden* or "Civil Disobedience," you have to remember that the Transcendentalists were Americans who were still within sight of the Revolutionary War—some of Washington's troops were still alive—and who thought that another societal revolution could still be achieved, if Americans only took control of their society, their government, their country. "Now the merchants in America occupy the place which was once held by the fighters and next by the nobles," wrote Theodore Parker.

IN THE FATE OF THE TRANSCENDENTALISTS lies the modern destiny of Thoreau. They are remembered for their Romanticism—Emerson's paean to nature, Thoreau's house-building exercise, the failed Brook Farm experiment, emphasis on failed. But what happened to the other side of the Transcendentalists? It is sometimes argued that their issues were lost to the one overriding issue of pre–Civil War America, slavery, which then consumed the nation for the subsequent decades. The stress on individual reform was, moreover, misinterpreted, in a sense, by the culture at large. What was initially an impulse toward a kind of humanitarian socialism that stressed individual rights was transformed, in relation to the expansion of capitalism in everyday life, into an every-man-for-himself philosophy that ended up being along the lines of what might be called democratic liberalism. You can follow Emerson's call for nonconformism all the way through the late twentieth century, when, as in a children's game of "telephone," it is mangled into the call for Americans bullied by market forces to take control of their own lives, *without* the help of government. "Let a man know his worth, and keep things under his feet. Let him not peep or steal, or skulk up and down with the air of a charity boy, a bastard, or an interloper in a world which exists for him," wrote Emerson, metaphorically. A century and a half later, Ronald Reagan told Americans to pull themselves up by their own bootstraps. The call changed from "Let's help ourselves to help each other" to "Let's help ourselves."

But to think of the Transcendentalist legacy as limited

to personal bootstraps is to emphasize the Trascendentalists' means at the expense of their hoped-for ends. Almost to a person, the Transcendentalists left their club of the like-minded—*disbanded* would be too strong a word because they were never actually together—and went deeper into the tasks of social reform. The paradox is that their emphasis on what they saw as a more enlightened and socially conscious religious experience, a religious experience that was simultaneously more personal and social, would end up sounding like an emphasis on personal religion—less like choosing a religion or a code of morals and more like choosing a personal creed and a pilates class. This left in the dust the Transcendentalist dream of a shared common humanity working toward a principled society, based on the radical justice they saw in the Christian principles they believed had faded in the Christian churches of the establishment.

They may be thought of as woods-wandering romantics today, but in their time, the Transcendentalists made a radical call for nonconformist reform—or nondogmatic experimentation, to use the critic Lawrence Buell's phrase. Reform was a trend, a fad, to some extent, and in the end, the Transcendentalists' call for reform was swallowed up by all the other calls. So that over the course of just a few decades, Theodore Parker's John F. Kennedy–esque "A Sermon for Merchants," of 1846—

> Remember your opportunities—such as no men ever had before. God and man alike call on you to do your duty. Elevate your calling still more; let its nobleness appear in you. Scorn a mean thing. Give the world more than you take. You are to serve the nation not it you; to build the

> church not make it a den of thieves, nor allow it to apologize for your crime, or sloth.

—eerily transforms into *How to Win Friends and Influence People,* by Dale Carnegie, who changed his name from Carnagey to Carnegie so that it might sound like that of the steel magnate:

> Get the other person to do what you want them to by arousing their desires.

SOCIOLOGISTS, LITERARY SCHOLARS, and even anthropologists will tell you that how a group of people see nature has a lot to do with how they feel about the world they are living in. The world that Thoreau was graduating into was characterized by flux and change. It's almost too easy to see parallels between our time and then. New communication systems had appeared, like the telegram that wired the country in a few years. The roaring stock market was about to crash. The gold rush in California that, as later happened in Silicon Valley, made people no one had heard of before very rich very quickly, at least for a while. There were all kinds of new cheaply printed newspapers—steam presses increased in speed every few years—and, like today's blogs, they were particularly brash, blatantly partisan. No one knew what to do with the youth of America, whose future was either in the factory or on the street—the paths to adulthood were no longer what they once had been, and the rough and rowdy subculture, the hip-hop generation of its time, was rising. When the Puritans had arrived, they had described nature as raw material, a waste in its wastedness, a place to be put to practical, virtuous use. Now it was becoming a refuge, if you could get to it—if you could make enough money to find the time to seek refuge, anyway.

To many observers, the political system seemed incapable of handling the problems of the day. Politics was changing, the parties seemingly mostly interested in spectacle as the avenue of access; within Jacksonian democracy, the average man (women, of course, could not yet vote, though the suffragist movement was less than a decade from launching in upstate New York) seemed to be involved in everything from caucuses to parades to beefsteak dinners; it was the beginning of politics as civic association, as political club, the dawn of the masses-oriented political event. On the one hand, the new politics of America made it seem as if the common man might run the show, implying a modernized democracy—a republic, but extra large. On the other, it meant that no one might have any control or, worse, that a few would have political power over the mass of men—a small elite might control the destiny of the nation, or at least the economy. As if to prove the point, President James Polk eventually used nationalist sympathies to start a war with Mexico that was illegal, a land grab that is sometimes called the United States' first elective war.*

At the same time, people's work lives were changing, partly as a result of the fact that they had a work life—before they had just had lives. Manufacturing was shifting from small artisans and crafts makers to factories. Young men went west for work, and women went to the nearby factory towns or, given the absence of men in New England, to jobs

*Polk had claimed American forces had been attacked in the United States, though they had not been, and a young and lanky politician from Illinois, Abraham Lincoln, who voted against the war, replied by saying, "Show me the spot!"

as teachers—the feminization of teaching happened in the years just prior to the Civil War. Just in terms of technological developments, the years around 1837 must have felt like before and after the Internet, when the notion and subsequent ability to "be online," as we now know it, changed everything from communications to shopping to dating.

The railroad was perhaps the most significant change. In 1830, when Thoreau was a boy, there were twenty-three miles of railroad track in the United States; by the time he died, there would be about thirty thousand miles of track. Obviously, the railroad changed transportation, but, more than that, all over America (as well as the world), the railroad changed the scale of things and, thus, people's perception of the world. It transformed the landscape the way the car would a hundred years later, and it transformed how people lived and interacted with one another and with their settlements, the way the interstates would starting in the 1960s. As the traveling time between places collapsed, the boundaries between cities and towns changed, and the populations of cities swelled. (And not just in the United States, of course: one million people lived in London in 1800; a hundred years later, six million lived there.) People began to think of distances in regard to shorter and shorter units of time, and of time in terms of schedules. Once, time was the provenance of the sun and the seasons; now factory shifts were as likely to determine a day, or what was being called a day. Before, everywhere had been local. Trains made everything seem more national, a trend that would eventually result in the idea of "global." In 1839, the *Quarterly Review* commented on the advance of the railroad: "As distances were thus annihilated, the surface of our country would as it were shrivel in size until it became not much bigger than one immense city."

Even the sensations of travel changed. Up to this point, people had traveled only as fast as their feet or a horse could carry them. Now they were approaching speeds of forty miles an hour. People no longer simply experienced places with all their senses, coming slowly into a town by foot or horse or wagon; people experienced a place visually, as a rush, as a scene from the window, later as a film. It was the beginning of the blurring of cities and towns, the visual merger that would eventually lead to the view along the suburban highway that we are familiar with today. The time I drove my family to Walden, for instance, we drove a little ways out of town and pretty quickly got on a many-laned local road bordered by strip malls, at which point my son awoke from a backseat nap and asked himself a question that he then answered: "Where are we? Oh, we're everywhere."

As people bought newspapers on their way to factory jobs and read about the world, as the new waves of displaced workers and immigrants began to fill the streets of Boston and as commuters—a new term for a new group—went in and out of Boston every day, as train tracks opened up new parts of the country, as the great western migration began, Thoreau was about to claim his home ground, a little pond. In the time when the world was feeling global, or at least national, he was going local. He was about to become a prophet of his hometown, when everyone else was racing to or from somewhere else, when the experience of your home had to do with its distance from another place.

It was a nerve-jangling time but it was also a hopeful time. The same events that were displacing people and drastically changing life in America might end up making for better ways of life—or at least faster and more efficient ways. Innovations in quicker and cheaper printing created the modern newspapers and, as Thoreau knew, the beginnings of an extensive American publishing industry, a business that had previously involved simply the rebinding of British book sheets. This would mean a flourishing of American letters, along with, for example, the birth of American pornography. And then there was the California gold rush. Karl Marx and Friedrich Engels, having completed the *Communist Manifesto,* wondered if the gold rush exempted the United States from the march of history; the immediate wealth (and intense liberty associated with it) was, Engels wrote to Marx, "not provided for in the Manifesto: the creation of large new markets out of nothing."

Thoreau and Emerson and the Transcendentalists hoped the turbulence of the times would propel change. They hoped that their modern world could reapply the same enlightened principles used in declaring independence sixty years earlier, the time of their great-grandparents. The Transcendentalists thought the idea of the United States of America was itself transcendental, and they wanted America to transcend itself one more time.

It is important to note that the first Transcendentalist that Thoreau spent any time with was not Emerson. Instead, it was the most vocal and even strident of the Transcendentalists, a man who was at an almost opposite side of the intellectual camp from Emerson. The Transcendentalists fell into roughly two categories—those who felt that society should bring about reform through laws, and those who felt that society could be reformed only when individuals themselves had undergone reform. It was a battle between the ideas of reforming from the outside and reforming from within. Each camp had in mind the same goal, a re-revolutionized society. Though Thoreau is easily placed on the side of the Transcendentalists who sought individual reform, he was no stranger to the other side of the Transcendentalist platform: Thoreau's first job outside of Concord involved living and working with one of the most vocal proponents of social reform, the Unitarian minister Orestes Brownson.

While Thoreau was at Harvard, the school passed a resolution permitting students to take an absence from the school for a few months, a ruling that was meant to allow low-income students who had difficulty affording tuition the ability to make money teaching. Thoreau found a job teaching in Canton, Massachusetts, working with Brownson. Brownson is as unremembered today as he was infamous in his day. Ralph Waldo Emerson was the refined spokesman for the Transcendentalists, the godfather of his Transcendentalist-poet-to-be, but the first Transcendentalist to influence Thoreau was Brownson, the philosophical ward heeler, the people's advocate, a loud, tough, and burly social reformer, who wrote, according to historian Arthur M. Schlesinger, Jr.'s biography, "the best study of the workings of society written by an American before the Civil War."

Brownson started out a Presbyterian, converted to Universalism, and then became an Owenist, supporting the New Workingman's Party, one of the first workers' parties in the United States. He proceeded on to Unitarianism, then to Transcendentalism, and, finally, disdaining the Transcendentalists with the same enthusiasm he had once used to help create them, to the Roman Catholic Church: Brownson was doing a theological one-eighty. In 1834, Brownson gave a speech in Dedham, Massachusetts, on the Fourth of July, arguing that the political equality espoused by the

Declaration of Independence had not yet been realized in terms of social equality. In his early days, Brownson was bent on overcoming the egotism and individualism that he believed characterized the age—Americans, he believed, mostly thought just about themselves, even if they joined reform movements and new churches and new associations and societies. He preached to young people, attempting to draw them into a "movement party," a party he felt stood for real change, a big theme then as well as now. Thoreau was one of the young people drawn to the movement and the idea of change.

The Transcendentalists can all be described as radical, but, in terms of pure economic analysis, Brownson was maybe the most radical. In 1840, in an essay entitled "The Laboring Classes," he suggested the destruction of the banks—which were, it seemed to him, more interested in making money than in circulating it—and the elimination of hereditary wealth, a severe impediment to economic justice. He described two sides to every village: first, the "neat and flourishing" side, where the people operating the new factories lived, and then, in a phrase reminiscent of the opening of Thoreau's *Walden,* the other side, where lived the mass of men, or in this case women:

> The great mass wear out their health, spirits and morals without becoming one whit better off than when they commenced labor. The bills of mortality in these factory villages are not striking, we admit, for the poor girls when they can toil no longer go home to die. The average life, working life we mean, of the girls that come to Lowell for instance, from Maine, New Hampshire, and Vermont, we have been assured, is only about three years. What becomes of them

> then? Few of them ever marry; fewer still ever return to their native places with reputations unimpaired. "She has worked in a factory," is almost enough to damn to infamy the most worthy and virtuous girl.

Your neighbor, Brownson noted, was the man profiting from their labor, as opposed to the laborers—"one of our city nabobs, reveling in luxury. Or he is a member of our legislature . . ."

At the time, the old-school Harvard Unitarians could argue that poverty was an enviable position, that the impoverished ought to be content, that hunger was like a condiment, affecting the flavor of the food in a positive manner. "The ploughman munches his mouldy crusts with as good a relish as the rich man eats his dainties with, for he has that best of sauces, hunger, to season his victuals," one of Thoreau's professors quoted approvingly, which gives you an idea of where the Transcendentalist irritation with entrenched religion came from. Brownson was a hardscrabble philosopher: "Now, the Great work for this age and the coming, is to raise up the laborer, and to realize in their own social arrangements and in the actual condition of all men, that equality between man and man, which God has established between the rights of one and those of another."

The reigning economic system, as Brownson saw it, put a greater value on some people over others. A majority of factory workers were women, women who might work from five in the morning until seven at night, who were often underpaid, who lived in company housing that was smaller and considerably less sanitary than the company agents' houses—women who were, as Brownson noted in "Laboring Classes," tossed out when they reached physical exhaustion.

These factories came out of nowhere, such that farmers were not pleased to see them arrive, fearing for their labor pool. Some towns did not want the squalor that was already known to be associated with factories in British manufacturing towns. Thus, the corporations tended to locate in relatively remote places that subsequently became central. In Lowell, in 1820, a group of businessmen set up a factory in a field where cows had been grazing. By 1840, it was a planned industrial site consisting of twenty thousand people. The point being that if you are thinking that Thoreau is living in a New England filled solely with nature walkers nature-walking, then you might want to try thinking about Thoreau living in a time of labor unrest, a time when his sisters had to break into a profession like teaching, which historically favored men, or go off to work in the mills popping up everywhere, while everywhere there were fiery speeches, workers' newspapers, and ministers trying to save the workers from exploitation, not to mention overstimulation. In this setting, Brownson was a rabble-rousing star.

Thoreau applied for a teaching position alongside Brownson and, beginning in the winter of 1836, spent six weeks in Canton, Massachusetts, a town with factories. Thoreau taught Brownson's children and, with their father, debated and conversed late into the evening, discussing the German writers who held sway over American intellectuals at the time. Thoreau saw himself as having a northern European ancestry, relating *Thor* to *Thoreau,* and in these early years, he was deeply impressed with the playwright and poet Goethe's ideas of an all-encompassing life design, a pattern that could be discovered anywhere, such as the veins of a leaf. It was Thoreau's first time in the constant company of an adult intellectual, and in this case he was not hanging out with a poet or a mystic. He was with a reformer, a radical street preacher. Thoreau wrote to Brownson afterward, describing their time together as "the morning of a new *Lebenstag*"—a new life.

Thoreau returned to Harvard from Canton, finished up his courses, and, between the end of classes and graduation, went to stay with a school friend in Lincoln, a town adjacent to Concord. They stayed about a mile east of Walden, in a little shack that the friend had built on the shores of Flint's Pond.

Chapter 4

A LIFE WITH PRINCIPLE

WHAT TO BE? What to do? How to make a living and at the same time live? These are the questions dangling before Thoreau when we first meet him, when the historical record of Thoreau kicks in, which is around the time of his graduation from college—when letters are saved, when people begin to write to him and about him and he writes to them, when he is spotted.

In Concord, or on the train between Boston and Concord, the Thoreau you might have seen was not very tall, and he was lanky, walking sprightly. His hair was brown, not unkempt but not terribly kempt. "His face, once seen, could not be forgotten," said his friend Ellery Channing. "The features were quite marked." Which is to say, he was on the homely side, though not at all uncharming. To

Channing, "quite marked" meant a large Roman nose—"like a beak, as was said." Thoreau's eyes were set deep, in a way that people seemed to think made him look more morose than thoughtful. His eyes were described variously as blue or gray or blue-gray, as mournful eyes that could flash with brightness when he was filled with enthusiasm, which was often. With a dark complexion, skin tan from all his time spent hiking around outdoors, some people said he reminded them of a sailor. He often walked with an umbrella; in his hat he built a little scaffold to hold small plant samples from his walks. He could set off on a trip to Worcester, for instance, with his supplies in a small sack or folded paper, and when he did, people would confuse him for a bum.

He was about halfway through his life at this point, and as he got older, he aged dramatically, his eyes seemingly deeper set, sadder, his nose bigger, more Roman, his skin more ruddy and, until the end, weathered, tan—the overall effect, according to a Concord neighbor, Franklin Sanborn, his first biographer and a contemporary, "reminds me of some shrewd and honest animal—some retired philosophical woodchuck or magnanimous fox. . . .

"He walks about," Sanborn continued, "with a brisk, rustic air, and never seems tired."

He was self-deprecating about his appearance, almost Lincolnesque. When he acquired some fame, when people wanted to meet him, he suggested the best of him was in his books, his works of art. "I am not worth seeing personally—the stuttering, blundering, clod-hopper that I am," he wrote.

He knew too that he was a character, and he seems to have been as unself-conscious about his body as he was self-conscious of it. He describes the scene he makes as he imitates a goose, honking and flailing his arms. Nathaniel Hawthorne's wife would laugh as he skated; Emerson was self-conscious and stiff in their three-writer skating parties, Hawthorne himself skittish, but Thoreau would twirl and spin and race around in a condition that was somewhere between a crash and a leap, a constant free fall.

HE WANTED TO BE A WRITER, a poet. He had just begun keeping the journal in which he would dig and toil and plow away for the second half of his life. If you are a writer and you want to write, then writing is one of those things you do whether you become known professionally as a writer or you don't. But writing *per se* was not really a career alternative for Thoreau. As is done today, he could contribute bits of writing here and there but people did not support themselves writing books in antebellum America, and to become a newspaper editor would have seemed a waste, given his education, and a step down, given his poetic goals. In 1837, a graduate of Harvard generally faced four professional options: a career in law or medicine, a life as a minister or a teaching post. Thoreau's father and older brother and sister were teachers. Thoreau became a teacher too. He managed to land a job straightaway at the Center School, Concord's public school. It was a difficult job to get and a difficult job. Thoreau was responsible for one hundred students. The rooms were in bad condition, alternately freezing or full of wood smoke; the students were continually fighting; supplies were less than limited. It was a good job because it paid well and was in his town—but also simply because it was a job. The United States had just entered an economic downturn—and then, starting with the Panic of 1837, would enter a severe depression.

The Panic of 1837 marked the end of a boom period begun in 1825. It was a result of speculation and the government's fiscal policy: after a large expansion of credit and loans and an expansion of the money supply, the wheat market failed or, in modern terms, the wheat bubble popped. As often happens, the effects rippled and led to a larger and more long-lasting disaster than a high school history book might let on. It was an economically disastrous time that imprinted itself firmly on Thoreau's professional mind-set—his papers and manuscripts are covered with jottings on accounts and money earned, as were the papers of Emerson, who, even though he was on a list of Massachusetts's richest men at one point in his career, was constantly chasing down payments due.

High prices, problems at the banks, and foreclosures based on out-of-control speculation: as we know too well today, it was a recipe for not just a Massachusetts depression. This depression ran through the country, affecting the old farmers and the new factory workers alike, with low farm prices, bread riots, rent riots, poverty around the nation. There were flour-trader riots in New York. George Templeton Strong, the New York City diarist, wrote at the time, "Failure upon failure . . . strong fears for the banks . . . and if they go political convulsion and revolution, I think, would follow." Cotton-trading companies began to fail, workers were laid off everywhere. Pennsylvania coal mines stopped mining. A Wall Street bank president apparently committed suicide, and the largest crowd of people ever assembled in Pennsylvania gathered outside Independence Hall to assail the banking system, which, many critics felt, allowed the speculating of those with money to the detriment of those who did not have a lot.

The economy, meanwhile, continued to undergo structural

changes, as the United States shifted from the age of the artisan to the age of manufacturing. Many factory workers ended up unemployed. Orestes Brownson compared slavery with working in a factory, citing the wage laborer as having "all the disadvantages of freedom and none of its blessings." Said Samuel K. Lothrop, a Unitarian minister, in 1837 in the *Christian Examiner*: "We were in the midst of peace, apparent prosperity, and progress when, after extensive individual failures, the astounding truth burst upon us like a thunderbolt. . . . That we were a nation of bankrupts, and a bankrupt nation."

The fact that Thoreau had a job must have made his family, at least, fairly happy. Likewise, it must have been a shock when, just a few weeks in, he quit.

After two weeks, he had been told by a school official that he was not flogging the children enough, a criticism he responded to, as the story goes, by flogging children at random and subsequently resigning. (This Thoreau story is probably apocryphal: the next school he worked at was his own, and he and his brother were apparently not altogether against corporal punishment.) The job paid $600 a year, a lot at the time. Then again, he was at that stage of life when you can sometimes find yourself telling people where to go, even if you are ill with tuberculosis and there are no other jobs to be found.

Shortly thereafter, Thoreau began a letter-writing campaign in an effort to land another job at a school or—his greatest hope, according to a letter to a friend—teaching a wealthy man's children: the kind of job that might allow him free time to compose. In the meantime, he got more involved in the life of his hometown, a lifelong trend, despite popular misconception. He was made the secretary of the Concord Lyceum, the local lecture hall, and around this time, he delivered his first lecture there, entitled "Society." He also made his debut as a published writer, his first piece appearing in the Concord *Yeoman*. It was an obituary for Anna Jones, an eighty-eight-year-old destitute spinster. "Poverty was her lot, but she possessed those virtues without which the rich are but poor," Thoreau said. It's easy to see the piece as a kind of public service, Thoreau decorously remembering an elder.

But in that time of flux, Anna Jones was, for Thoreau, a bridge to the era of the generation of the Revolution. She was a contemporary ancient. "After a youth passed amid scenes of turmoil and war, she has lingered thus long amongst us a bright sample of the Revolutionary woman," he wrote. "She was as it were, a connecting link between the past and the present—a precious relic of days which the man and patriot would not willingly forget."

Up until his graduation from college, the town records still listed his name as David Henry Thoreau, but around this time he officially changed it to Henry David. At home it was no big deal; his family had always referred to him as Henry David. But the change would eventually irritate the people who were irritated by him, as it still does today.

"His name ain't no more Henry D. Thoreau than my name is Henry D. Thoreau," said a farmer years later. "And everybody knows it, and he knows it. His name's Da-a-vid Henry and it ain't never been nothing but Da-a-vid Henry. And he knows that!"

Just as he would irk Emerson by spending more and more time with the farmers in town as his time in Concord went on, so he irritated farmers when he did things that seemed to them to be overly scholarly or pretentious. He was the townie with airs at the same time that he was the intellectual who was actually a townie—a switch hitter, or at least reversible, like his name. Thoreau can easily seem too full of himself, and yet he can have the same complaint of others, making him a Rorschach test in the end.

"I don't like people who are too good for this world," he wrote in his journal, just after college, in that too-good-for-the-world postgraduation tone. "Let a man reserve a good appetite for his pack of dirt."

That was when he was first journalizing, just after he met Ralph Waldo Emerson.

IMAGINE MEETING EMERSON. Imagine being a young Harvard grad and discovering that the preeminent public intellectual, the writer everyone is talking about, lives in your hometown. Imagine he had moved there and written the essay "Nature," a work that threw down the intellectual gauntlet—as well as the literary gauntlet, given that Emerson was, for the most part, a writer. He was a rock star in his day, in fact, though his hits were ideas, the chief idea being reform, the chief hope for reform being in the youth of America—in the person he called the American Scholar, a type of person who could easily have been a recent Harvard grad living in Concord. There was a transatlantic mood, a feeling that America was in the midst of developing its own arts and literature, and Emerson was the public point man. His essay "The American Scholar" has sometimes been referred to as the nation's intellectual declaration of independence.

In manner, in appearance, and in dress, Emerson was elegant and refined where Thoreau was homely and coarse. Six feet tall with a Roman nose which, in Emerson's case was a note of elegance, Emerson's cosmopolitanism contrasted with Thoreau's deep-rooted localism. But today, Emerson, like Thoreau, is often misremembered. He can be written off as the first American Dead White Male, or his words can be used to support an argument that is, on review, anti-Emersonian. As I have already noted, he is singled out, mostly, as the thinker who created the national myth of rugged individualism, who spawned the Marlboro Man as the characteristic American citizen, the self-sufficient go-it-alone-er.

Like Thoreau's, Emerson's sentences can act as workers for hire, temporary employees that can be turned against their creator and themselves. Quoting from Emerson's best-known essays can paint him as the opposite of what he was contrasting, can paint him, in other words, as a one-note thinker most often taking the small view of the individual: "A foolish consistency is the hobgoblin of little minds," to use one of his most famous lines.

An example: At the 150th anniversary of Emerson's birth, a *New York Times* editorial showed him to be a posthumous supporter of modern Republican tax cuts for the wealthy, callously indifferent to the down-and-out. The editorialist cited "Self-Reliance": "I tell thee, thou foolish philanthropist, that I grudge the dollar, the dime, the cent I give to such men as do not belong to me and to whom I do not belong." The point here is not that he was against giving. The men who "do not belong" to him are the dime-a-dozen philanthropists, the "thousand-fold Relief Societies." On the contrary, he asked people to find the best way to give, asked them to look at the mechanisms of charity. He cautioned against using what we conceive as our goodness as a varnish coating what in ourselves is mean.

Emerson was living in a time of revivalists, of causes in

general—of, as he notes, meetinghouses built for meetings that never took place. Antebellum America was a nation where the machinery of philanthropy and do-gooding, even the pursuit of it, began to feel empty. Emerson was the preacher who left the church, declaring its institutions spiritually bankrupt; he was against, not the aim of the church, but rather its methods. He was against, not organizations or even conforming, only "conforming to purposes that are dead to you." He was called a mystic and he was, but, like his student-to-be, he was not a mystic in the visionary sense. "He was not seeking in the angle of vision an escape from the world; as it formed, the angle of vision was to make 'use' of the world," wrote the literary scholar Sherman Paul.

Emerson was a reformer among reformers in a time of reform, and he was, in a number of instances, addressing methods of reform. Abolitionists were everywhere in New England, but slavery still existed. In "Self-Reliance," Emerson chides the abolitionist who welcomes the freeing of slaves in the Caribbean—in England, the Slavery Abolition Act of 1833 ended slavery in the British Empire in 1834—but shows little or no interest in those enslaved in various ways in his or her own community. "Go love thy infant; love thy wood-chopper; be good-natured and modest; have that grace and never varnish your hard, uncharitable ambition with this incredible tenderness for black folk a thousand miles off."

The Emerson who met Thoreau was a cautious enthusiast. He was simultaneously a radical *and* a conservative, egging you on and holding you back. The son of William Emerson, the well-known Unitarian minister at Boston's First Church, he was reluctant to embrace the family business. He was ordained after much pondering, and then resigned after three years. His first wife, Ellen Tucker, died a year and a half after they were married, of tuberculosis, and then one brother died and then another—death envelops the Transcendentalists. He remarried, to Lidian Jackson, but for all intents and purposes, it was, as he described it, a practical marriage: "This lady is a person of noble character whom to see is to respect," he wrote to his remaining brother. He dreamed of literature, and of starting a magazine; and on a doctor's advice that a longer trip might help him with an obstinate case of diarrhea, he chose a visit to Europe over the trip he was about to make to the West Indies and, sailing on Christmas Day, 1832, immersed himself in European Romanticism, setting out to meet William Wordsworth and Thomas Carlyle.

He came to Transcendentalism after Transcendentalism as a movement had already begun. "Nature," his seminal essay, was loved by the other Transcendentalists not so much for what it said as for how it said it. "At this point in his career, his chief charisma derived from a challenge to conventional wisdom rather than for his particular wisdom itself," Philip Gura writes. What the essay said was that nature was man's starting point, nature being the stuff of everyday life, that which we use to build our homes and cities and grow food, all the physical world taken at once. The link between nature and us was, for him, language. To the Transcendentalists, language was connected to the real world in a primal, precivilized way: words were rooted to natural facts. It was thus through language that we had access to the divine mind; with etymological digging, links to divine reason could be discovered, at the moments of words' beginnings. This interest in language also had a pop expression: like many Americans, the Transcendentalists were obsessed by word roots, a fad at the time, like crosswords, or sudoku today. In the case of the Transcendentalists, the difference between a connection to the Logos and a pun was the difference between yoga as spiritual excercise and yoga for exercise.

Empiricism ruled the day—facts reigned, and science was out to explain everything—but this was not enough for Emerson, who sought a mystical and romantic transcendence in relation to nature. In his mind, nature, as opposed to organized religion or social institutions of any kind, was what we ought to use to calibrate our lives: the nature all around us, the nature of which we were constructed. To put it in contemporary electronic terms, nature is your hard drive, and you don't need a church or any human institution other than yourself to start it up or, more relevantly, to reboot. Reform and true sight requires a cultivated awareness of the life around you, everywhere around you—what he called "the pot-luck of the day," a great phrase—as opposed to what we think of when we think of nature today, which is in the park with the trees. For a person living, say, in a big city today, it's the ratty-and-partially-green-potpourri-of-life-around-you version of nature rather than the Sierra-Club-calendar-photo version, and you have to bond with it, even when it is less than extraordinary.

"The invariable mark of wisdom," Emerson wrote, "is to see the miraculous in the common."

EMERSON WAS, by most accounts, a truly transcendental speaker. Rather than as a philosopher, he came across as something more like a philosophical poet; Elizabeth Peabody called "Nature" a prose poem. Orestes Brownson recognized Emerson's rock star–like charisma early, telling his Unitarian superiors that they had to pay attention to the issues Emerson addressed "or to Ralph Waldo Emerson they may rest assured their pupils will resort." Though he critiqued associations, he was himself a mainstay of the then-developing lyceum movement, which, like book clubs and subscription lecture series, spoke to people's demand for self-improvement. He could fill lecture halls, as he often did on tours throughout New England, and eventually into the western states and territories. Emerson was no social-justice radical, pounding a pulpit of reform. But he was riled up, being born in the nation-forming time of Thomas Jefferson and John Adams, and living now under James Polk and Franklin Pierce, when the nation seemed, not just to not have matured, but to be about to grind to a halt. The rise of politics as mass culture was vile to the Whig in him: the immigrants, factories, robber-barons-to-be were "rude, lame, unmade . . ." He could not see collective action as the answer to the ills threatening American society, and as a result he was attacked by the likes of Orestes Brownson, who thought that Emerson's self-culture could not restore men's inalienable rights once they had been stripped away in the marketplace: sure, you could get in touch with the divine that seeped through the world, but what did that matter if you were worked to death and poor?

But Emerson was radical in that he worked so hard to avoid fixed opinions; he was stubborn about the idea that you ought to change your mind. "Whoso would be a man must be a nonconformist," Emerson says, and he proves it with his nonconforming sentences, rhetorical incitements that also stress how difficult what he asks of us is for him—as his relationship with Thoreau over the years shows. He is the anti-leader, the guy who is supposed to be influencing but who is, paradoxically, charged up by the fact that he is so influenced by everybody else: "Every decent and well-spoken individual affects and sways me more than is right."

Like John Stuart Mill, Emerson saw the strength of the individual as the key to the strength of society, and he was set on preaching to one person at a time, in his essays, in his books, on his lecture circuits. The thirst for public education was part of what led to his increasing popularity. He was, even after rejecting his religion, a preacher, self-stripped of his own pulpit, delivering not essays but what he called "lay sermons." And to some extent he loved being out there—how else can anyone stand a life on the road? Emerson's self-culture was by no means antisocial. He is a modern figure in that he wanted to work alone, while he longed for company. Besides, even the Transcendentalists who were predisposed toward personal reform, as opposed to social reform, were all too concerned about the world to turn away from it—they were highfalutin gadflies. In "Society and Solitude," Emerson calls for a balance: live your life in society in sync with the revelations that you discover in your personal solitude. Get it together, then come back, which is to say, the team is better if each player is in better shape. It's important to not let what everyone else is saying intimidate you: "We require such a solitude as shall hold us to its revelations when we are in the street and in palaces; for most men are cowed in society, and say good things to you in private, but will not stand to them in public." He admits the difficulty, the tendency to be driven off course: "These wonderful horses need to be driven by fine hands."

Thoreau's mentor is a contradiction; he reads like one of his own essays, and they are easy to misread, taking left turns to eventually end up going right. He believed in Margaret Fuller, for instance, and cherished her brilliance (and wrote her letters that, if he had been married in the way most people are married, might have concerned his wife), but at the same time he could talk about the protofeminist as if she were a ditz, sounding ditzy himself in so doing. He was late to speak out for abolitionism in the United States, and for this you want to shake the guy, pull his head out of the sand he has so carefully raked. But as humans, our great strengths are often also our greatest weaknesses, and eventually Emerson did get behind his abolitionist friends—he changed his mind. And because he was Emerson, the nation's first and foremost public intellectual, because he was, in a way, the time's one and only talking head, he brought a lot of people with him.

As the Emerson we know, Ralph Waldo Emerson is a kind of selfish patrician who created the antisocial hermit Thoreau. The Emerson we don't know is the thinker who tests everything, who is skeptical of even his own stuff, who attempts to appreciate and even encourages people he doesn't really appreciate—Thoreau, for example. Emerson is the person who wants to break out of the box even if he is sitting in one. "His central project," writes Lawrence Buell, "was to unchain human minds."

IMAGINE THOREAU, FRESH FROM SCHOOL. Imagine him standing on the doorstep of his future, infused with the Greeks, translating Romans, ready to do battle with the world. Imagine him ready to strike out, his Transcendental "I" at its most first-personal. Having already spent time with the most vociferous, most prominent Transcendentalist—the gruff and comparatively street-savvy Orestes Brownson—Thoreau, as he graduated, was now to meet the movement's great stylist, the bad boy of religious and intellectual circles, circles that at the time were close to the same circle. Emerson gave the address at Thoreau's commencement that would later be published as "The American Scholar." (It is possible that Thoreau was not there, that he was hanging out back in Concord, but Thoreau would have read it.) "The scholar is that man who must take up into himself all the ability of the time, all the contributions of the past, all the hopes of the future," he said. "He must be a university of knowledges," Emerson said.

The address was a call for the life of the mind, a plea against the anti-intellectualism that characterized American culture then as now. And yet it's not all about books: it says that experience and education in the real world should be valued as well as what is learned in books. (You can see how upset Emerson will be later on, when his star pupil chooses to live in Concord as opposed to the larger American world, even though, at the same time, Emerson kind of wants him to stick around.) It says that the world is nature, that resource seemingly abundant in America. To a certain American artist at the time—and even later, if you look at Ralph Ellison, for example—Emerson is the coach in the locker room: Whitman reads Emerson and takes his work as if it is a handoff, a signal to run the ball up the poetic field. To Thoreau, Emerson is calling him into the game, calling for a new kind of literature—a kind of writing that comes from looking at yourself and nature. "And, in fine, the ancient precept, 'Know thyself,' and the modern precept, 'Study nature,' become at last one maxim," said Emerson. It was a call to a poetic artistry that reforms, that changes souls and thus the world.

How did they meet exactly? It's not certain. But Concord was small. Emerson had moved there in 1835. Emerson's sister was a boarder at the Thoreaus' house, and at one point Emerson's sister passed one of young Thoreau's poems on to her brother, and then Emerson had, of course, heard about the poem that Thoreau had tied up with a bouquet of wildflowers and tossed through the window of Mrs. Lucy Brown, who had been a boarder at the Thoreau house, along with her two sons. Some Thoreauvians seem to think Thoreau had a thing for older women, and he may have, though in this case, given his role as local handyman, he was just as likely to write Mrs. Brown a poem as fix her stove:

I am a parcel of vain strivings tied
By a chance bond together,
Dangling this way and that, their links
Were made so loose and wide,
Methinks,
For milder weather.

He was indeed a parcel of vain strivings, as Emerson knew. Emerson called him a "brave, fine youth," and was quickly suggesting topics and even lending him cash—$100 to begin his search for a teaching job. Less often noted is the unfailing consistency with which Thoreau paid back Emerson and that Thoreau very rarely executed the writing assignments in the manner that Emerson suggested. An example is Thoreau's journal. Emerson suggested it, but Thoreau executed it with an intensity that Emerson really couldn't understand until well after Thoreau died. Its famous opening begins with a question that is usually ascribed to Emerson:

> "What are you doing now?" he asked. "Do you keep a journal?" So I make my first entry today.

Chapter 5

A FREE-LANCE

FOR A LOT OF WRITERS, writing is something that has to be worked through, like a bad cold; the goal is not to write but to think. As Thoreau set out to find a job, his journal was his main literary occupation, and in the beginning it was stiff and self-conscious. It was *writing*, as opposed to the more effortless *mind thinking* that marked his later entries, when he made lists and charts and day after day of straight observation, when things he saw were close to things he thought. In the fall of 1837, though, the journal pages were filled with youthful wonderings, short treatise-like entries on human nature, brief passages packed with meaningful intention, with headings such as DISCIPLINE, HARMONY, THOUGHTS, and his first entry, SOLITUDE:

> To be alone I find it necessary to escape the present—I avoid myself. How could I be alone in the Roman emperor's chamber of mirrors? I seek a garret. The spiders must not be disturbed, nor the floor swept, nor the lumber arranged.

But Concord is in the early journal, for example, during the fair: "As I pass along the streets of the village on the day of our annual fair—when the leaves strew the ground, I see how the trees keep just such a holiday all the year." His journal would always be his sounding board, his life written out, and the struggle of the early days was to weigh the competing needs of solitude and society:

> *A talent is builded in solitude,*
> *A character on the stream of the world.*
> *He only fears man who knows him not, and he who avoids him will soonest misapprehend him.*

As he added those first pages he was mostly getting the hang of journalizing, and seeing the journal as a workplace, as opposed to what it will become in a few years, an end in itself. "My Journal is that of me which would else spill over and run to waste,—gleanings from the field which in action I reap."

As he looked for a teaching or tutoring job, he made himself useful, something he did all through his life, with odd jobs, letter writing, and always reading book after book. His reputation today is as a loafer, but he was a most productive loafer. He wrote to his brother John, who had moved out of town for work, suggesting they both head west to find teaching jobs, which would have made for a whole other Thoreau—a Thoreau on the Prairie. "It's high season to start," he wrote to John. "The canals are now open, and traveling comparatively cheap. There's nothing like trying." He helped Emerson try to find a house in Concord for their friend Margaret Fuller, and Emerson in turn lent him $10 to take a boat to Maine to look for a teaching job. Arriving in Portland, he is stiff, restrained, feeling jostled and, probably, homesick: "A sensitive person can hardly elbow his way boldly, laughing and talking, into a strange town, without experiencing some twinges of conscience, as when he has treated a stranger with too much familiarity." In bustling waterfront Maine, he is the reticent poet, the refined college grad, who sounds perhaps like the Thoreau you know—shy, unresponsive, wanting to keep to himself. On the other hand, he manages some elbowing, despite his misgivings, and when he travels farther down east and lands in Oldtown, he meets an Indian, and begins a lifelong obsession with Native Americans, whom he decides might really have something to say. "He was the most communicative man I met," Thoreau noted.

In May 1838, Thoreau returned home from Maine jobless. He advertised in the local paper for a school of his own. Emerson saw all this as an admirable avoidance of the Boston rat race. "My brave Henry here who is content to live now, and feels no shame in not studying any profession, for he does not postpone his life, but lives already—pours contempt on these crybabies of routine and Boston." This is Emerson overanalyzing, not realizing that Thoreau and his brother were wondering about the West, where the rats were racing to in 1838.

He took long walks with Emerson, to the nearby cliffs

and hills, to Walden Pond and its environs especially. Emerson was falling in love with walking at this point, for a number of reasons: he was emulating Wordsworth, a woods walker, and looking for solace after the death of his brother Charles, and Thoreau was his woods-walking mentor. About to go off on a hike through the nearby woods with a group of women, Emerson called for Thoreau, asking him to bring his flute—"for the echo's sake." After a walk, they attended a concert together given in Concord by a Swiss group, the Ranier Family, a hot ticket touring the country to great acclaim beginning in 1839. The Raniers were a Tyrolean family band that, with their tour, inspired families all over the United States to form family bands and sing of their mountain roots, even if the American family bands didn't actually have mountain roots and just pretended they were from the hills. Another popular band was the Hutchinsons, who later became known as the New Hampshire Rainers; their song "The Old Granite State" opened with the line, "We have come from the mountains." English ballad-singing was popular in people's homes, and sheet-music ballads were the MP3 downloads of their day. Thoreau loved singing ballads. "Will You Come to My Mountain Home?" was a hit at the time, meaning it sold a lot of copies as sheet music. It was written by Francis Henry Brown, who was one of the first successful American composers without European tutelage—he was brought up on hymns and fiddle tunes in Newburyport, Massachusetts, where he was born in 1818, a contemporary of Thoreau.

It's difficult today to imagine how much music was a part of regular social life at this time, how much music meant to everyone—Thoreau loved music, something you learn when you read his journals but not when you read books about him. Thoreau loved the Raniers, and described aspects of the concert for days in his notes. "When I hear music, I flutter, and am the sense of life, as a fleet of merchantmen when the wind rises," he said. He also mentioned that he began to laugh at their feather hats and *lederhosen* until he saw how serious the musicians were, at which point he stifled his laugh, to be polite.

EMERSON WAS WHAT MIGHT BE described as infatuated with Thoreau, as was his wont. Emerson was an extroverted introvert, a private person who dramatically and publicly embraced ideas and people. Thoreau was, by contrast, an introverted extrovert, by default quiet on the outside, but itching on the inside to be public. These are relationships that are partly intense friendships—as was the case, on the one hand, with Emerson's relationship with Margaret Fuller—but that also feel more than vaguely homoerotic, especially when compared to his sometimes unfeeling, or at least dispassionate, letters to his second wife, Lidian.

Deciphering the Emerson-Thoreau relationship on the most intimate level—and, further, deciphering the intimate relationships between men in their time—was and is complicated. For instance, it is difficult to define their relationships in terms of homosexual and heterosexual relationships as they are commonly defined today, mostly because close, even romantic relationships between men, like those between women, were not unusual at the time. In the 1800s, there was no word for homosexuality. Brothers grew up sharing beds. College roommates often did too, to save money. (As a young lawyer, Abraham Lincoln rented a room and part of a bed in New Salem, Illinois, from a storeowner named Joshua

Speed, who, when he proposed it to Lincoln, did not know him at all.) "From time to time, a close male friendship in youth would blossom into something more intimate and intense," E. Anthony Rotundo wrote, in an essay entitled "Romantic Friendship: Male Intimacy and Middle-Class Youth in the Northern United States, 1800–1900." "Warmth turned into tender attachment, and fondness became romance. An ardor developed between young men that would seem unusual outside of gay circles in the twentieth century." It's difficult to say exactly how prevalent these kinds of romantic relationships were between young middle-class men, but they required no apologies and apparently did not cause condemnations, a way in which pre–Civil War America was slightly less uptight than the America today. These close relationships generally ended when the men married women. Women were more likely to keep their same-sex relationships going throughout their life.

Scholars who see Emerson as being gay or bisexual point first to Emerson's infatuation during his junior and senior year at Harvard with a student, Martin Gay, known as "Cool Gay," who was in his thoughts, Emerson noted, a "dozen times by day and as often by night." But that analysis on its own misses the mark. On the one hand, his obsession with Gay was more like a crush, as Emerson himself admitted, and ended after graduation. On the other hand, the definition of male friendships was much broader.

I'll return to the question of Thoreau's sexuality a little further on; it's an issue that's important to keep in mind, it seems to me, when Thoreau's legacy begins to take shape in the years after his death. But for now let me just say that Emerson liked Thoreau a lot. He was determined to help Thoreau start a literary career of his own and to help keep him afloat financially. While writing his own addresses, and beginning a new Transcendental journal, the *Dial,* Emerson paid Thoreau for chores. At the same time, he advised Thoreau on reading, suggested topics for lectures and essays, wrote letters of introduction, and, as always, went to Walden Pond to hear him play the flute, even before Thoreau moved there.

IN MANY BIOGRAPHIES AND CRITICISMS, Thoreau has been pinned as a mere influence of Emerson's. Henry James called Thoreau "a translation of Emerson into the sounds and fields of the forest." But James was not seeing the extent to which Emerson relied on Thoreau. Over the years, even during periods of strain in their friendship, Thoreau was Emerson's assistant and second-in-command, doing carpentry, surveying, and gardening at the Emerson house, teaching and minding the Emerson children. He was a classically trained handyman, who could build a room on Emerson's barn when he wasn't translating Greek poetry for the *Dial.* In 1859, at the death of Emerson's mentally handicapped brother, Bulkeley, who had been living with a farmer in Littleton, Massachusetts, Thoreau was asked to make all the funeral arrangements, as he had with Emerson's mother. When he went on trips and lecture tours, Emerson, who was a constant and careful reviser, entrusted Thoreau with the printer's galleys of his books; Thoreau was in charge of final corrections.

Emerson served as an intellectual springboard for Thoreau and vice versa. In some ways, Thoreau taught Emerson how to live in Concord. Thoreau was envied as the local Renaissance man, the person who could, for instance, "pace sixteen rods more accurately than another man could measure them with rod and chain," Emerson noted. Emerson said this as a compliment, but tucked in with the compliment was a nod to class differences: Emerson could afford to hire Thoreau, and not vice versa. Emerson came from a patrician family of ministers; Thoreau's father was a merchant struggling to make a living. The middle class was born in the years before the Civil War, and one of the ways it began to define itself was by separating itself from manual labor, including even household chores.

While there was much reciprocity, Emerson often treated Thoreau as a servant, or what was becoming known as a servant, when Thoreau wanted mostly to be associated with the new kind of work. In a rhetorical shift, writers of the time began to associate work with actions that had not been considered work before, like writing. Hawthorne, for example, talked about his "intellectual labor." At the same time, working with money began to be considered working with one's hands. Farmwork, on the other hand, began to be associated with health and exercise—at least voluntary farmwork or farmwork on a farm that was not a primary source of income. As patricians, or what we might today call upwardly mobile people, separated themselves from physical labor, writers and reformers and social commentators began to associate manual labor with relaxation and a new concept, exercise. Like today, those who could afford to choose chose manual labor because they wanted to, not because they had to. Sports took on greater importance, as affirmation of the value of competitiveness and for health purposes. The more people like Emerson shunned manual labor in their labors, the more they wanted it in the rest of their life. In 1833, the Society for Promoting Manual Labor in Literary Institutions promoted manual labor schools, which involved wood chopping, craft making, and fieldwork. The president of the association (using upper-case letters for emphasis) suggested that manual labor schools "WOULD TEND TO DO AWAY WITH THOSE ABSURD DISTINCTIONS IN SOCIETY WHICH MAKE THE OCCUPATION OF AN INDIVIDUAL THE STANDARD OF HIS WORTH."

In large part because of these labor issues, utopias were the rage, though they weren't called utopias. They were called associations. Both Thoreau and Emerson were offered a spot at the Brook Farm experimental community being established by their friend George Ripley and his wife, Sophia, a group including a still relatively unknown story writer named Nathaniel Hawthorne. There, they would share the work evenly; class distinctions would theoretically be ignored. Thoreau's rejection of the offer is often thought of as emblematic of his antisocial proclivities. "As for these communities," Thoreau wrote to himself, "I had rather keep bachelor's hall in hell than go to board in heaven." This line (from his journal) is often dredged up as proof that he could not live with people. But he was not against the communitarian aspect of Brook Farm. As opposed to Emerson, he found almost all work attractive. He was being asked to live in a boardinghouse, and he already lived in a boardinghouse—his parents' home. Starting out, he was playing with the idea of getting away from *that* life, even though he never would, and ultimately wouldn't want to; heaven was on earth for Thoreau, anyway. He was already boarding in hell, and he could handle it. Thoreau was quicker to respond to Ripley than Emerson, who characteristically hemmed and hawed and in the end decided he couldn't live in a cooperative community like Brook Farm.

Ultimately, even in the area of work, Thoreau was both influenced by and influencer of Emerson, as Thoreau's mother pointed out when one of her boarders commented on resemblances between Emerson and her son—she had always seen her son rubbing off on his older friend.

"O yes," she said. "Mr. Emerson had been a good deal with [my] Henry, and it was very natural he should catch his ways."

As Emerson said in his Harvard Divinity School address in 1838, "Truly speaking, it is not instruction, but provocation that I can receive from another soul."

So imagine the thirty-four-year-old Emerson meeting this young Concordian scholar, this promising poet who knew nature, at least the nature that was Concord and its field and stream environs. Imagine the promise that the Concord Sage saw in the Concord Pan. Imagine the sheer entertainment value—for Thoreau was, true to his modern reputation, an arguer, a twenty-two-year-old who relished an opposite. At first, Emerson loved this about Thoreau, though later it would drive him crazy, and he would peg Thoreau as bitter even though the bitterness might be coming from Emerson. "He wanted a fallacy to expose, a blunder to pillory," wrote Emerson.

Said Thoreau, "Resistance is a very wholesome and delicious morsel at times."

Resisting the Brook Farmers may also have had to do with Thoreau wanting to fend for himself as a farmer—something that perplexed Emerson, that seemed to him a resistance. A few weeks before Ripley's invitation, Thoreau had asked a man about renting some land. "[H]e said he had four acres as good soil 'as any out doors,'" he wrote in his journal. Either way, Thoreau was enjoying being in Concord, especially given what started out as a utopian friendship with Emerson.

And as he talked about writing with his mentor, often taking the opposite, he began to see what was natural in what is merely human. He began to see nature as a way to write about human nature and—most important—about society, in this case, the society of Concord and America. "One must needs climb a hill to know what a world he inhabits," he notes in his journal, in the fall of 1837.

THOREAU CAME FROM A FAMILY OF TEACHERS, and it seems inevitable that he would end up teaching again, even though he never altogether dropped the idea of farming. After returning from Maine, he started his own school. He had just four students but shortly the enrollment grew to eight, so that he soon wrote to his older brother, who was working as a teacher out of town, to convince him to return to Concord to help him. Thoreau, perhaps in reaction to his experience as a student, chose to make learning be about the positive experience of learning, the joy. "I would make education a pleasant thing both to the teacher and the scholar," Thoreau wrote in a letter to Orestes Brownson. "This discipline, which we allow to be the end of life, should not be one in the schoolroom, and another in the street. We should seek to be fellow students with the pupil, and we should learn of, as well as with him, if we would be most helpful to him."

Henry taught Latin, Greek, French, and physics; John taught math and English. They worked to engage the children, rather than feed them information by rote. The two brothers took the students on long walks in the woods; recess was thirty minutes instead of ten. They opened the windows for fresh air, a radical innovation. In addition to visits to Native American archaeological sites—what we would call a field trip today—the brothers took the boys to local businesses, the printing press and the gun shop. When their boat needed tarring, they took them along to help. They learned how to survey Fair Haven Hill, a field trip that served to introduce Thoreau himself to the craft of surveying. By night the teaching brothers were, according to a neighbor, "comrades in all the parties, dances, and various schemes of entertainment which occupied the leisure hours of the young people of the village." By night, the two brothers were, in other words, partying.

Teaching by morning, strolling the fields in the afternoons, reading Greek, and then socializing by night: not bad, as rent-paying work goes, though teaching never sunk in as a profession for Thoreau. He considered himself a poet. He was now leaning more toward the career of his mentor, Emerson, who was publishing essays and books and making a living as a lecturer, though a dicey one; Emerson's letters are full of financial frettings, and he is often hitting the road to make money—"off peddling," as he called it. (The road is the lecture circuit, in this case.) Emerson encouraged Thoreau as a poet, though even at that time there was very little chance to make a living as a poet, unless you were, like Henry Wadsworth Longfellow, teaching as well. Thoreau enjoyed the coarseness of Geoffrey Chaucer and played with both obscure and popular poetic forms of the seventeenth and eighteenth centuries. More and more he wrote about his environs, marrying the long-lost epic with the prosaic in his local scene:

If with fancy unfurled
You leave your abode,
You may go round the world
By the Old Marlborough Road.

Eventually, he sent his poems and other writings to the *Dial*. Emerson had helped found the *Dial*, but Margaret Fuller was the editor, and, to his credit, Emerson respected her decisions. Fuller accepted some of Thoreau's poetry, but not his essays, which he was especially anxious to publish. Every writer hopes for an editorial path of least resistance, and Thoreau must have thought of the *Dial* as a sure thing, given his relationship with Emerson. But Emerson, always on the road, was more of an editorial director, and Thoreau was stuck with Fuller, who, aside from being tough, was brilliant, and ultimately right to reject Thoreau's early essays, despite Emerson's persistent nudging.

"This essay is rich in thoughts, and I should be *pained* not to meet it again," Fuller wrote to Thoreau, about an essay entitled "The Service." "But then the thoughts seem to me so out of their natural order, that I cannot read it through without pain."

He had worked on the piece for a year; she had held on to it for a tortuous five months before deciding to reject it in such an acutely artful manner. How he must have wished that he did not think her brilliant! She encouraged him to work on it more and submit it again, which, to his credit as a tireless craftsman, he did. Fuller was his toughest editor, and though he wrote some good poems, he had her to thank in part for accidentally steering him into his hybrid prose making, sentences that were poetry filled, lyric inspired. He was at that stage in his craft when confidence exceeds skill, when the writer feels his writing mostly shines, and it mostly doesn't.

In Thoreau's life, as in most everyone's, the things he chooses not to do are as important as the things he chooses to do. Around this time, he decided *not* to get married. In the summer of 1839, at twenty-two, Thoreau met a girl, Ellen Sewall, a friend of the Thoreau family and the seventeen-year-old sister of one of his favorite pupils. He had already been infatuated with Ellen's younger brother, composing a poem, "Friendship," about him. "There is no remedy for love but to love more," Thoreau wrote in his journal. Meanwhile, John Thoreau, his brother, had also fallen in love with Sewall. This complicated matters significantly; it was a nightmare scenario for Thoreau, his best friend in love with his heart's desire. They seemed to court her simultaneously for a while—rowing on Walden Pond, walking, picnicking, and horseback riding—but as soon as John expressed his interest, Thoreau fell back. At that moment, the two brothers took off for their school's summer recess on a boat trip to New Hampshire. Thoreau was, by all indications, in shock, emotionally speaking—his concern, as he noted in his journal, "source enough for all the elegies that ever were written." He didn't bring up the matter, it seems safe to say, with anyone; like one of his botanical specimens, he pressed his love deep in his heart. He must have imagined Ellen would be smitten with John for all the reasons that everyone else was. He must have felt doomed, in that John was the outgoing Thoreau and he was the quiet one. The account of their New Hampshire trip would eventually be the basis of Thoreau's book, *A Week on the Concord and Merrimack Rivers*. As a book, it is a patchwork of Transcendental meditations, woven into a brief narrative of the boating excursion, but if it had been a reality TV show, it would have surely been a lot about the fact that Thoreau probably wanted to kill his brother for falling in love with his sweetheart.

Not that Thoreau would have brought it up on TV, even in one of those interviews that, on reality shows, are recorded off set. In his journal he is secretive. "I cannot make a disclosure," he writes. You wonder why he didn't call the New Hampshire trip off, but they left as planned. Just before they launched, Thoreau threw a big melon party, a yearly melon party that was becoming his trademark in Concord among the people who liked him and, given the quality of the melons he grew, possibly among the people who did not like him so much. He covered a long table with the various kinds of melons he relished cultivating and, by at least one account, served them with wine.

After that, the two young men boated for four days to Hooksett, New Hampshire. They walked and then took a stagecoach into the White Mountains, where they hiked to Franconia Notch. Upon turning around, after recovering their boat, they caught an incredible wind that blew them back to Concord in a day, racing down the river as if in a transcendental dream.

When they returned, John headed to Scituate to visit Ellen. Henry must have been miserable for those few days. The next week, the brothers traveled together to visit Ellen. Upon their return to Concord, John sent Sewall opals for her rock collection. Henry sent her father a book of what were essentially religiously racy poems—which was some kind of message, for sure, and perhaps not very smart, as far as winning over the father goes. Ellen's father was an uptight Unitarian minister who considered himself of solid New England stock and thought both Thoreaus too Transcendental for his daughter. Nevertheless, the following summer, John proposed. You can hear the gulp in Thoreau's journal. Ellen accepted, though her mother quickly convinced her to break it off, for her father's sake. That fall, Henry tried to marry her, proposing by letter. He was immediately rebuffed by Ellen, also by letter, on the advice of her father, who immediately shipped Ellen away to upstate New York, for protection from the Thoreau brothers. "I never felt so badly at sending a letter in my life," Ellen confessed to a friend.

Thoreau rarely used his journal to detail his most personal thoughts, and in this case he plowed himself into his writing until the crisis passed. To assuage his sadness, he took longer and longer walks in the woods, and though he was suffering setbacks as a writer and now had been rejected in love, he managed to see the fields and woods as beautiful, as joy giving. This is his trademark emotional turn: when he is down, he meditates, burrows in sadness, in deep, dark thought, then comes back up through the earth happy, newly resolved, a crocus breaking through snow.

"Nature refuses to sympathize with our sorrow," he wrote in his journal, a few days after he is rejected by Ellen. "How all trees tell of the sun," he wrote.

You could say he sublimated his emotions into the natural scene around him, which is along the lines of what a famous psychological profile of Thoreau from the 1970s argued—Richard Lebeaux's "Young Man Thoreau." Lebeaux sees Thoreau as having had an identity crisis of an Oedipal nature. Or you could say that his emotions had opened up the way your sinuses open up when you take an antihistamine, or just get outside and breathe again. Either way, he would not propose to any woman ever again, and very soon his life would become a kind of priesthood, devoted to his writerly goals, and in devotion to his brother, still his best friend, who was about to die.

THOUGH IT WAS SUCCESSFUL, the Thoreau brothers' school was forced to close after two years when John was hit with a bout of tuberculosis in 1841. Thoreau was happy to have more time to write, but he was again jobless. He would have liked to write full-time, but he had to make a living, as well as contribute to the family's finances—even today, rare is the writer who can merely write as he pleases, financially speaking. As usual, he toyed with the idea of becoming a farmer, and met with several farmers about their property, even though he did not necessarily have the money to buy a farm. Emerson invited Margaret Fuller for a visit, hoping she and Thoreau would get to know each other editorially and that, as a result, Fuller would be more inclined to publish Thoreau. Fuller, to Emerson's chagrin, liked the idea of Thoreau being a farmer, and encouraged him in an agricultural direction. Thoreau was again reading Virgil's *Georgics,* the account of the seasons that seems on the surface to be a book about farming, but is about stewardship and life, a work of poetic realism, that, for Thoreau, pointed to the profit of sticking close to home. "It was for [Thoreau] the great poem of the Earth," wrote Robert Richardson in his intellectual biography of Thoreau, *Henry Thoreau: A Life of the Mind.* Richardson argues that the *Georgics* had stylistic relevance to Thoreau as well as geographic relevance. "It is a cornucopia of vivid detail and plain example, and the agricultural world it describes is remarkably similar to that of the American farmers Thoreau knew." Work, in the *Georgics*, is espoused not as a burden but as a means to earn and enjoy good things.

Once again, Thoreau wasn't sure what to do with his life. He castigated himself for not being on a more certain path, while at the same time marveling in the profit from tasks both literary and nonliterary. He had been digging manure, for instance. "Great thoughts hallow any labor," he wrote. "To-day I earned seventy five cents heaving manure out of a pen, and made a good bargain of it[.]" (Hawthorne comes to the opposite conclusion at the farm collective: that digging manure is a waste of his time.) Thoreau began to talk about living on a pond again, as he had immediately after college. He joked that his employment would be simply watching the seasons. Then, he changed his mind, and solved his problems—for the moment, anyway—by moving into Emerson's house, as the unhired man.

The move satisfied a coincidence of interests for Emerson and Thoreau. With labor reform in mind, the Transcendentalists, and Emerson in particular, were uncomfortable with the idea of hiring servants—housekeepers and babysitters and manual laborers. Emerson spoke of America's "servant problem." Rapid industrialization had separated the home and workplace, making waged workers in the home seem suddenly strange. Liberal New Englanders felt guilty about paying people to serve them; southern slavery apologists poked fun at them, arguing that southern slaves were more cherished than northern domestics, given that domestics might be fired at any time. "We love our slaves," wrote George Fitzhugh, a social theorist who authored the pamphlet *Slavery Justified!* Fitzhugh was egging on northern abolitionists, and his rhetoric affected northern liberals because of their conflicted feelings over servants, feelings that, of course, still exist today.

As part of his and Lidian Emerson's one-year attempt at

what Emerson called "labor & plain living," a "family-style" domestic servant situation, Emerson encouraged the cook and the maid to eat at the same table as his family. Lidian called Emerson's plan "a wild scheme." (The maid would eat with them, but she eventually stopped because the cook could not be convinced to leave the kitchen.) With the coming and going of so many visiting Transcendentalists—what Lidian called the "Transcendentalist Times"—Emerson seems to have nearly run down his staff, one of whom, probably in Thoreau's earshot, threatened to post a sign: "This House is not a Hotel." Emerson was, in other words, rather clueless, as well as a rather choosy democrat. The communal aspect of the Brook Farmers appealed to Emerson, but not the farm aspect. Rather than go away to a communal living place, he was trying to create a communal living place at home, and thus encouraging his friends to live in Concord. Thoreau would be a combination of *au pair* and editorial assistant, a literary apprentice who took the kids on walks, as Emerson helped him in his literary pursuits. It was a sort of domestic literary barter.

"We work together by day in my garden, and I grow well and strong," Emerson wrote to Thomas Carlyle. After work, they took care of leisure by taking to the woods, excursions that, I would hasten to add, given that the Concord-based Transcendentalists were all busy scrounging for cash, were cheap—free-lance writers tend to do a lot of walking. "Through one field only we went to the boat, and then left all time, all science, all history, behind us, and entered into Nature with one stroke of a paddle." Thoreau planned to stay one year at the Emersons' but ended up staying two. His room was at the top of the stairs, and he had the full run of Emerson's library. The tendency is to think of him as always monkish, but there he was doing housework and chasing down the kids, or entertaining them with tricks or walks to the pond and frogs—and then trying to bang out a poem during the kids' nap time. (One of his arsenal of child-amusing stunts: pencils would disappear and then be drawn from the children's ears.) Yet he was already going a little nuts in the first few weeks. He had room and board and extra pay, but he was not striking out on his own. He took advantage of Emerson's library but could feel like his mentor's copyboy, lesser by comparison, demeaned. Thoreau referred to the arrangement as "a dangerous prosperity." But the prosperity ended that first winter when his brother died.

DEATH LINGERED ON THE EDGES of life in Concord; Margaret Fuller commented that young men who did not go west often died before they could contribute fully to the community, often of tuberculosis, with which Thoreau was already battling. But this death was sudden and unexpected. While shaving, John Thoreau cut off the tip of a finger. He replaced the bit of skin and wrapped it in a bandage. A few days later he was suffering from lockjaw, his body stiff like a board. There were violent muscle spasms, his neck bending back, his body arching ferociously. The doctor who came from Boston told the Thoreaus they could only wait for the end. Ten days later, on January 11, John died at the age of twenty-seven—in the arms, it is said, of his brother.

In the days following, Thoreau seemed to his family oddly calm, uninterested in the outdoors even, and then almost acquiescent, until a week or so passed, at which point he took strangely ill. On January 24, Emerson wrote to his brother William, in New York, that Henry, who had been staying at his father's house, had lockjaw. By all accounts, Thoreau suffered the same skeleton-tightening muscle spasms that had made his brother's death so horrific. The doctor diagnosed him with lockjaw, though he did not appear to have an infection of any kind. "It is strange, unaccountable, yet the symptoms seemed & on the increase," Emerson wrote.

Thoreau himself would later describe it as a case of "sympathetic lockjaw."

Then, just as Thoreau recovered, Emerson's five-year-old son, Waldo, died of scarlatina. Thoreau's journal was silent for weeks. He wrote letters expressing his grief to Lucy Brown, the older woman in Concord whom he still occasionally sent poems to, and to Isaiah Williams, a young man who had come through town seeking Transcendental advice (already Concord was a mecca). Thoreau tried to console Emerson, with Stoic thoughts. "How plain that death is only the phenomenon of the individual or class. Nature does not recognize it, she finds her own again under new forms without loss. Yet death is beautiful when seen to be a law and not an accident," Emerson wrote to Thoreau from New York, where Emerson had gone on business. Most biographers think Thoreau never got over his brother's death, but then why would he? When you read his journals, the sadness returns almost every winter at this time, the way a river at a certain time in the fall turns up its bottom.

Some people handle depression badly; it cripples them, saps the energy out of their life and body. For other people, it melts the ice and allows them to look deeper. In Thoreau's case, the death of his brother melted something inside him. It revitalized him, or vitalized him, depending on how you look at it. Writing to Emerson, he cites death as the source of life, and to himself, he sees it as a challenge: months before in his journal he had thought about the death of a close friend. "On the death of a friend we should consider that the fates through confidence have devolved on us the task of a double living—that we have henceforth to fulfill the promise of our friend's life also, in our own, to the world." Now it seemed as if he was taking his own advice. Now he took on the role of John, the sociable Thoreau, the public Thoreau, modulating his earlier reticence about public taste, about lowering himself, as a writer, to please public taste. At this point, the journals are newly charged, and, as this new Thoreau emerges from the cocoon of the old, he uncharacteristically addresses himself to God. "Why did you include me in your great scheme? Will ye not make me a partner at last?"

If before he had sometimes felt a need to hold back in some way—"I shall hold the nobler part at least out of service," he wrote months before—now he discusses his need to "give," "to serve the public."

"If I should help infuse some life and heart into society—should I not do it a service?" he wrote in his journal.

It was a stroke of editorial good luck for Thoreau that, in March of 1842, at the very same time the newly purposed Thoreau was emerging, Margaret Fuller gave over the editorship of the *Dial* to Emerson—every once in a while, things break the right way for a free-lance writer. The *Dial* was dying and only had a few issues to live. "Poor *Dial*!" Emerson wrote. "It has not pleased any mortal." But for its final issues, Thoreau would be assistant editor, publishing poems, translations, and, at last, a long essay, "The Natural History of Massachusetts." He went from being a talked-about player on the bench to the utility man, covering all the bases. Emerson handed him a series of scientific reports and surveys by the state of Massachusetts and asked him to write an essay. It was partly for Thoreau's sake: "As private secretary to the President of the Dial, his work and fame may go out into all the lands, and, as happens to great Premiers, quite extinguish the titular master," Emerson wrote to a friend. And partly for the sake of the *Dial*: pieces on natural history were all the rage in the popular press, as Emerson, who rose to fame on an address entitled "Nature," knew well. A very similar essay had run, for example, in the *Knickerbocker*—"Natural Science . . . Lessons of Town and Country," ostensibly reviewing the same Massachusetts reports. And there was Gilbert White's "Natural History of Selborne," one of Thoreau's favorite works. With his very first piece of so-called nature writing, he was staying within the genre, not trying to create anything new, ostensibly.

The essay was as Emerson set it up to be—Thoreau's own thoughts on the nature of Massachusetts, or mostly Concord and environs, using the reports as a jumping-off point:

> We were thinking how we might best celebrate the good deed which the State of Massachusetts has done, in procuring the Scientific Survey of the Commonwealth, whose result is recorded in these volumes, when we found a near neighbor and friend of ours, dear also to the Muses, a native and an inhabitant of the town of Concord, who readily undertook to give us such comments as he had made on these books, and, better still, notes of his own conversation with nature in the woods and waters of this town. With all thankfulness we begged our friend to lay down the oar and fishing line, which none can handle better, and assume the pen, that Isaak Walton and White of Selborne might not want a

> successor, nor the fair meadows, to which we also have owed a home and the happiness of many years, their poet.

Thoreau then heartily assumed the pen from Emerson. Like Walton and White, the English exemplars of nature writing, he starts out as the genial tour guide, a couch naturalist; he begins the piece pleasingly, not at all didactically, and without a hint of the Transcendentalism that will be laced throughout:

> Books of natural history make the most cheerful winter reading. I read in Audubon with a thrill of delight, when the snow covers the ground, of the magnolia, and the Florida keys, and their warm sea breezes; of the fence-rail and the cotton-tree, and the migrations of the rice-bird; of the breaking up of winter in Labrador, and the melting of the snow on the forks of the Missouri; and owe an accession of health to these reminiscences of luxuriant nature.

It was a different kind of piece for Thoreau, as well as for the *Dial,* in that it was heavy on the nature and light on the theology. It was genre writing, to a large extent, with a Transcendentalist at the wheel. The Transcendental ideas were there—"to help infuse some life and heart into society," as Thoreau had written—but they were subtler, nearly coded. The reports are the subject of the essay, but Thoreau abandons the subject immediately, and soon the narrator is on the trail of a fox, reworking his own journal entry on a fox spotted in the Concord woods and smoothly shifting back and forth between first and second person, luring the reader in. As he follows the creature, he traces the very idea of the fox from storybook fox to real fox, and then back to the fresh trail that suddenly becomes a trail of a thought, of a creature's mind thinking, hints of the divine in the natural world: "I tread in the steps of the fox that has gone before me by some hours, or which perhaps I have started, with such a tiptoe of expectation, as if I were on the trail of the Spirit itself which resides in the wood, and expected soon to catch it in its lair." This Thoreau, a more vivid, more exciting, less cerebral writer, still presents himself as the one "brave" individual, as he did in the essay Margaret Fuller rejected on service. Here he suggests the notion of looking at nature as the health-restoring balm to civilized man, "a silent and unambitious" cure for the "din of civilization."

Margaret Fuller, naturally, did not like the essay. Emerson liked Thoreau's essay enough to print it, but just barely. "I do not like his piece very well but I admire his perennial threatening attitude, just as we like to go under an overhanging precipice," he wrote. The essay was not so threatening to Emerson as Thoreau was, this possibly too humorous poet, this trickster with words, and here you can hear the beginning of the rift between Emerson and Thoreau. Meanwhile, Thoreau resented Emerson not telling him the truth—that is, that he did not like it. Emerson preferred another piece in the *Dial,* Charles King Newcomb's "The Two Dolans," a piece that, unless you are Emerson, is incomprehensible, though even he praised it backhandedly as "the maddest piece in the issue." This is what Thoreau was up against in Emerson's circle: the Transcendentalists preferred aural ramblings, truth stumbled into, as opposed to truth taken through drafts, which was Thoreau's method, what the Transcendentalists referred to as "suicide of the pen." A writer intent on writing and rewriting was an anathema to them. They were obsessed with conversation. Alcott's digressive "Orphic Sayings" was more to Emerson's taste, though ironically, Alcott singled out only Thoreau's "Natural History" for praise in that issue of the *Dial*. Alcott could seem nuts to people, but he was more than sane enough to know that Thoreau wasn't.

Despite Emerson's reticence, "Natural History" was a breakthrough piece of writing for Thoreau. It was, as noted, his first nature essay. It wasn't his finest piece of writing; the cobbled together pieces of his journal make it feel like cobbled together pieces. But it showed his potential to be a captivating reporter on the fields and streams, and now as opposed to being known as a struggling poet with a penchant toward classical themes, he very suddenly had a literary reputation as someone who knew his way around the woods. Most important, he had a reputation, a writer for hire—the *Dial* was small but, mostly because of Emerson, well read. The *New York Tribune* called the piece "excellent" and described Thoreau as "a political philosopher," which must have made him very happy. Even the Boston *Morning Post,* usually unimpressed by the Transcendentalists, called it "a very good article."

And it won him a new friend. As a result of the publication of "Natural History," one of Thoreau's biggest fans was now Nathaniel Hawthorne, an up-and-coming romance writer who had recently moved to Concord—Thoreau had planted him a garden as a welcome-to-town gift—and after an initial meeting, where Hawthorne, Emerson, and Thoreau stood around not knowing what to say to each other, they all became friends, especially Hawthorne and Thoreau. Thoreau would take Hawthorne out in his boat on the river, and sometimes rescue Hawthorne from embarrassing social situations—once when people dropped by for tea, he slipped out the back door to meet Thoreau. Hawthorne was the reclusive one. (Thoreau eventually sold Hawthorne his boat, since he needed the money.) In his journal, Hawthorne described Thoreau's essay as "so true, minute, and literal in observation, yet giving the spirit as well as the letter of what he sees, even as a lake reflects its wooded banks, showing every leaf, yet giving the wild beauty of every scene."

If Emerson was now the encouraging critic, pushing Thoreau to change his style, Hawthorne more wholeheartedly encouraged Thoreau, possibly inspiring him to write up a travel narrative, another style of essay that was in demand by the periodicals that were then emerging as the primary publishing vehicle for American authors. Thoreau was already working on an account of his trip to the White Mountains with his brother John when Hawthorne mentioned the idea. Hawthorne was even willing to share his contacts. When magazine editors called on Hawthorne in Concord, he was quick to recommend Thoreau. "He writes; and sometimes—often for aught I know—very very well indeed. He is somewhat tinctured with Transcendentalism; but I think him capable of becoming a very valuable contributor to your Magazine."

Chapter 6

WHEN THE WOODS BURNED

SMELLING A CHANCE, the newborn nature writer was quick to write up his trip to Wachusett, a small but popular mountain just outside of Concord, drafting the story on the back of his late brother's nature album. He sent it straightaway to the *Boston Miscellany of Literature and Fashion,* on Hawthorne's recommendation. They took it. After all Margaret Fuller's hemming and hawing, after Emerson's reluctant enthusiasm for his pupil, Thoreau must have finally felt like a pro, especially because the *Miscellany* was known to pay well—the day a writer sells his first big piece is a day he remembers for a long time. "A Walk to Wachusett" is not his greatest work, by any means, and yet it is a miniature template for what is to come: he uses a form of popular literature to make his point, quietly, and often with humor. This

Thoreau was the opposite of the obtuse Thoreau, the obscure and inscrutable writer who disdained public interest. This Thoreau wanted a hit. In fact, the piece was even less Transcendental than "The Natural History of Massachusetts," on the surface, at least, beginning with its form. That it had any form at all was itself anti-Emersonian. (Emerson worked hard to write formlessly, inspired by intuition, which worked well for him.) But here it was a simple travel narrative, a walk.

"A Walk" announces itself as a shaggy dog tale, exclaiming its anticlimax in the introduction. Homer, as Thoreau says, "conducts his reader over the plain, and along the resounding sea, though it be but to the tent of Achilles." It's a joke, but on the other hand, earthly distance is not his goal: "In the spaces of thought are the reaches of land and water, where men go and come. The landscape lies far and fair within, and the deepest thinker is the farthest traveled." It was a trip into the west, a direction with classical significance to Thoreau—into the west is a trip to the past for the writer interested in the ancients. The west also had nationalist overtones: the west was the patriotic direction for Americans, the direction of America's Manifest Destiny. He is becoming a writer of opposites, and this story is, like all of his travel writing, anti-travel. The faraway tone of adventure is also belied by the fact that it is local. Wachusett, a mere two thousand feet high, is only about thirty-five miles from Concord, and they walked the distance.

Throughout his life, the mountains Thoreau climbed (with the prominent exception of Maine's Katahdin) were most often within a day or two's walking distance. In fact, despite his walking reputation, he often took a train. Wachusett is the hill on the edge of Concord. It was not a daring height, and the trek was not particularly strenuous; very rarely is Thoreau an adventure traveler. He covers land habitually covered and re-accesses it, and re-values it (though eventually comes to the conclusion that the value of land is incalculable). In Thoreau's time Americans feared their landscape in general was devoid of the ancient associations that Europeans could recognize. "Genius has not consecrated our mountains, making them high places from which the mind may see the horizon of thought widening and expanding around, over past ages," Sarah J. Hale wrote in *Traits of American Life,* in 1835, ". . . They are nothing but huge piles of earth and rocks, covered with blighted firs and fern." (In the 1600s, European writers said the same thing about the Alps.) Thoreau was out to consecrate the unconsecrated hills. Edward Hoagland, in an introduction to a book of Thoreau's writing on mountains, makes the point that Thoreau was user-friendly in his expedition choice, as opposed to, say, Shackelton, the South Pole explorer. "We can all be with him on a Sunday walk of an overnighter as long as we recognize, too, there in our tent, that Thoreau was attempting to recalibrate civilization, as well as simply sleep out. Nature was to stand in a central position, not as a vestige, or pet, or vehicle for our narcissism, such as it is when we scramble up mainly for an aerobic exercise to condition the body, like a stint of sidewalk jogging or pedaling gymnasium machinery."

On the walk to Wachusett, Thoreau recognizes points of American history ("This, it will be remembered, was the scene of Mrs. Rowlandson's capture, and of other events in the Indian wars . . ."), but for the most part he recasts the vista: "This part of our route . . . may remind the traveller of Italy, and the South of France." He is again applying the lesson of the *Georgics* to his home provinces: life is ancient and modern, and human nature is the river that runs through it. Most significantly, Thoreau speaks rhapsodically about the *built* landscape in the same manner as the *unbuilt* one; land that is touched by human is no less important than untilled land. The two exist peaceably, so that our great nature writer in his formative stage is reveling in the sounds of not just songbirds and streamlets but guns and cows: "Every rail, every farmhouse, seen dimly in the twilight, every tinkling sound told of peace and purity, and we moved happily along the dank roads . . . But anon, the sound of the mower's rifle was heard in the fields, and this, too, mingled with the lowing kine."

In the piece, the stylings of the more mature Thoreau appear like buds in winter—first, in that the essay takes the conventional travel genre and tweaks it slightly, almost cryptically. In contemporary terms, it's as if he has written a story about weight loss and has sneaked in a few words on spiritual heft. As the undervalued writer on the Transcendentalist team, he can get work only in the non-Transcendentalist world, and then only if he keeps his philosophical messages cryptic. It wasn't the first time the idea had occurred to him; in a college essay he had written that a writer publishing in the popular press had more influence than a preacher in a pulpit. But it was the first time he had tried it, and it would be his MO for many writings—for *Walden,* for instance. Still not quite accepted by Emerson and company, and still not trusted by the editors of the popular press who feared his Transcendentalist tendencies, Thoreau became a writer who was in no camp completely and, as such, eventually learned to write for two audiences simultaneously, the popular press and a reader he imagined to be like himself, who reads obsessively and is always thirsty for spiritual renewal. He learned to write in a kind of code, simultaneously pleasing himself and, he hoped, everybody else.

THROUGHOUT "A WALK," THOREAU MAINTAINS a scrupulousness of detail that surely heartened Hawthorne while irritating Emerson, neatly describing the towns and villages and views; as he approaches Mount Wachusett, the hill becomes "less ethereal," he jokes. He recounts the stays in farmers' barns, meals, meetings with farmers, and when he sits on the summit, at last, he sees a fire off on Mount Monadnock, "which lighted up the whole western horizon, and by making us aware of a community of mountains, made our position less solitary." In his new public spiritedness, he feels community rather than separateness and, in the course of saying so, manufactures one of his trademark aphorisms, a gem concerning the sparkling heavens: "Truly the stars were given for a consolation to man."

Most travelers in Thoreau's time, as in ours, would head to the mountains for restorative purposes, returning revived, ready to re-immerse. In "A Walk to Wachusett," Thoreau suggests not that we leave the city in the country, but that we bring the rural back to the city. He suggests a reverse spiritual commute, and in a style that mixes the voice of the man of leisure with that of the critic, a sugarcoated Diogenes. In the end, Thoreau's narrator comes back, not rejuvenated, as would be the case in the nature or travel piece of the day, but inspired to rejuvenate life at home, in town. In "A Walk," he suggests bringing the mountain back to the plains, to re-rural the urban. In so doing, life as usual is replaced, he suggests, by a new life:

> And now that we have returned to the desultory life of the plain, let us endeavor to import a little of that mountain grandeur into it. We will remember within what walls we lie, and understand that this level life too has its summit, and why from the mountain top the deepest valleys have a tinge of blue; that there is elevation in every hour, as no part of the earth is so low that the heavens may not be seen from it, and we have only to stand on the summit of our hour to command an uninterrupted horizon.

"A WALK TO WACHUSETT" WAS A SECOND HIT and, with "The Natural History of Massachusetts," a back-to-back score, and thus seemed to bode a change for the young writer. The publishers of the *Boston Miscellany* promoted "A Walk to Wachusett" vigorously in the New York and Boston papers, always a heartening sign to an author, and the *New York Tribune* reviewed the issue, singling out Thoreau's piece as the best. Unfortunately, after one more issue, the *Boston Miscellany* closed. Thoreau chased down his payment for months, enlisting Emerson's aid, as well as the aid of Elizabeth Peabody, the bookstore owner and publisher. The Transcendentalists had been a little squeamish about being paid for writing to begin with, but they got over that pretty quickly. "Instead of this scruple let them make filthy lucre beautiful by its just expenditure," Emerson justified.

In Concord, Thoreau helped Emerson set up a public reading room, and he spent a season as director of the lyceum, arranging for twenty-five speakers, and staying just under the $100 budget. There was a controversy when he invited Wendell Phillips, an abolitionist, to lecture, but Thoreau put Wendell's invitation to a vote, and Phillips was allowed. Thoreau's own lecture was on Walter Raleigh—he was steering clear of controversial issues, such as slavery, to draw a larger crowd, following Emerson's model. Emerson was away—he was $900 dollars in debt in 1843 and was off to lecture in Baltimore, Philadelphia, Manhattan, Brooklyn, and Newark—but his wife wrote him and praised Thoreau's performance. "Henry ought to be known as a man who can give a Lecture," Lidian told her husband. Even John Keyes, a prominent Concordian who had opposed Wendell Phillips's lecture on abolitionism, praised Thoreau, which especially pleased the budding lecturer and rhetorical trickster. He had won over a political enemy, a goal at this outward-reaching stage of his life.

While Emerson was away, Thoreau worked on what would be the last two issues of the *Dial,* proofreading and translating verses for Emerson, choosing sayings of Confucius and three of his own poems and a brief essay for inclusion. In the press, he was praised for his introduction to his own translations of Anacreon, the Greek lyric poet, while the translations were deemed so-so themselves; the *New York Tribune*—namely its editor, Horace Greeley, who was already a fan of Thoreau from afar—saw in Thoreau a gift for making ancient writings accessible for the modern reader. Part of the reason for this acclaim was that Thoreau, as a writer, was becoming a specialist in magazinese, the language with which magazines even today dress up their stories for sale: "We lately met with an old volume from a London bookshop," Thoreau wrote by way of introduction. (In fact, he hadn't lately met with the volume by chance, but had translated the lines as long as four years earlier.)

With his successes, Thoreau was becoming antsy in Concord, and on March 1, 1842, he wrote to Emerson, who was in New York again, suggesting that he too might go to New

York—Thoreau said he was "meditating some other method of paying debts than by lectures and writing," as he put it. "If anything of that 'other' sort should come to your ears in New York will you remember it for me." Emerson thought Thoreau was not ambitious enough; he was frustrated with the stubbornness and independent streak of his disciple, even though in another light he could applaud it. Emerson jumped at the chance to find Thoreau a situation. He arranged for Thoreau to work for his brother, William Emerson, a judge who lived on Staten Island. Thoreau would tutor William's son, Willie, and be supplied with room, board, and wood, and time and a room for writing. Thoreau agreed.

He was to begin right away, though he appears to have become ill, another bout of tuberculosis, and could not leave until May 1. While he waited, he got cold feet. As his friends and neighbors' hopes for him increased, so did his fears. "I am glad, on Mr. Thoreau's own account, that he is going away," wrote Hawthorne, "as he is physically out of health, and morally and intellectually, seems not to have found exactly the guiding clue; and in all these respects, he may be benefited by his removal." (Note the degree to which Thoreau's illness characterizes him, even in his twenties.) But Hawthorne added: "On my account, I should like to have him remain here, he being one of the persons, I think, with whom to hold intercourse is like hearing the wind among the boughs of a forest-tree; and with all his wild freedom, there is high and classic cultivation in him too." He was given parting gifts by a number of friends, and he in turn gave Hawthorne his music box to take care of. Emerson was to miss Thoreau's editorial skills, but Ellery Channing, Thoreau's poet friend and a mutual friend of Emerson's, joked in a letter to Margaret Fuller that Emerson already had a new young poet that he would now mentor instead, Benjamin West Ball, in place of that "oakum brained Thoreau."

CONTRARY TO ALMOST EVERYTHING you will read about Thoreau, he did not hate New York City. He did not like being away from home, and after a while, he may have even hated being away. But in the end, he managed to make the best of what was—in view of his artistic goals—a very bad move. By a number of measures typically discounted, he succeeded; the trip helped him see the city, the ocean, and, mostly, his home, as viewed from far away. He was already becoming a different kind of writer from what either Emerson or the other Transcendentalists at the *Dial* thought he might be and, in Emerson's case, thought he should be. New York locked in those changes, and they helped him see Concord as his main subject, even if he thought he might come up with a new subject, something more far-ranging. As Thoreau left Concord, Emerson wrote ahead to his brother: "And now goes our brave youth into the new house, the new connexion, the new City." (Emerson once wrote that, under different circumstances, he could imagine himself living in New York City.)

In hindsight, it might be said that the spring of 1843 was probably one of the worst times in history to try to break into New York publishing, not that it is ever a good time. It was certainly the right place to try to break into publishing in general: the industry centered itself there prior to the Civil War, as did just about everything else, it seemed to some people. "New York is becoming more literary," Longfellow wrote. "It will soon be the center of everything in the country,—the Great Metropolis." The publishing industry itself was expanding quickly, both a propagator and a product of the industrial revolution; Thoreau was developing as a writer as America developed as a literary marketplace, since cheap printing was advancing simultaneously. Stereotyping and electrotyping replaced setting type by hand (at least after a first edition), and the flatbed steam press and cylinder press were designed for speed. The expanding reach of the railroad increased distribution between the 1840s and 1850s. And there was a willing audience for books and literature in general: between the time Thoreau graduated from college and the beginning of the Civil War, literacy rates in the United States were the highest in the world. Basic literacy in the United States was at ninety percent for white adults compared to about sixty percent in Britain. (Literacy rates were higher for men than for women, and higher in the northeastern states.)

Thoreau faced both a high demand for writing and a glut. In the same way that modern-day publishers have attempted to republish writing on the Internet, magazines and newspapers in Thoreau's time largely reprinted foreign material, as copyright laws did not protect foreign publishers. Weekly magazines reprinted entire novels cheaply. Newspapers reprinted material from journals and magazines—which survived on subscriptions only, advertising not yet being in vogue—and were likely to be surviving one year and on the verge of bankruptcy the next, with subscriptions not renewed. "As to writing for the magazines, that is very nearly done for as a matter of profit," said the best-known magazine writer of the day, Nathaniel Parker Willis, when writing to a wannabe writer. The *Boston Miscellany* thrived when it did in part because it ran so-called fashion plates, which were used to print illustrations of women's fashion. These appealed to the growing number of women working in factories, for instance, in Lowell and Nashua and all around Boston, and were the disdain of intellectuals, such as Thoreau's comrades on the *Dial,* circulation roughly two hundred. *The Ladies' Companion* had a circulation said to be close to twenty thousand and published short tales and poems, both of which, as a result, became popular forms for writers and readers. It is still sometimes the case today that behind a successful freelance writer is a women's magazine.

The mechanics of paying writers had not been worked out, not that it ever will be. Longfellow had tried to find some work for a friend at around the same time, writing to publishers on his friend's behalf, then reporting back to the friend: "I got an answer, showing that nobody pays now-a-days," he said, and he quoted a publisher's note: "The fact is that all our publishers *whether of books or of periodicals, are desperately poor at present. Money is not to be had.*" A poor writer from a not very wealthy family looking to get into a business that was not at the time paying many writers made for a Thoreau who could sound fairly desperate. Impressively, despite his illness, despite his artistic predilections, despite the odds being stacked against him, he pressed on, letting no publisher go unturned, using humor and perseverance. "But I see that I must get a few dollars together presently to manure my roots," he wrote to one editor. "Is your journal willing to pay anything, provided it likes an article well enough?"

He was sick again, upon arrival, with a severe cold and bronchitis, and what is thought to have been a form of narcolepsy—he fell asleep in the evenings when he tried to write, a problem not just for writers who have young children—and his bad cold was underscored by the tuberculosis. He arrived at the docks in lower Manhattan in May of 1843 and, amid the flood of immigrants arriving, found his way to the ferry to Staten Island. He did not care much for this alternate Emerson family; the family conversation was dull compared to the Concord Emersons' table, and he did not take to his young charge, Willie. He liked the ocean, the salt marshes, and the hourly ferry, but the situation quickly brought him down. His first letters were to his family: he sent his sister a poem that he wrote about his brother's death, and he sent his mother a letter saying that, while he was homesick, he planned to make the best of it. He wrote Emerson's wife, Lidian, who was also sick at the time. He went a little overboard trying to comfort her, with compliments that sound very love-letter-y: Thoreau could make up for his outward stiffness with deep outpourings of epistolary friendship. "The very crickets here seem to chirp around me as they did not before," he wrote.

There is nothing like pulling up a writer to get him thinking about his roots. Distance sharpens the view, and it is not surprising to see that in these months, in New York City and its environs, Thoreau began to work up the sentences that are the foundations of *Walden*. Ironically, in those autumnal New York months, the sentences were about spring. Not surprisingly, they were often about Concord. "I have hardly begun," he wrote in a letter he would send to Concord, "to live on Staten Island yet; but like the man who, when forbidden to tread on English ground, carried Scottish ground in his boots, I carry Concord dust in my boots and in my hat,—and am I not made of Concord dust?"

HE KEPT BANGING ON THE PUBLISHERS' DOORS. He tried the *New Mirror,* the *New World, Brother Jonathan,* the *Knickerbocker* (which paid writers based on their reputation), and Harper Brothers, the book publishers, who said they were doing fine, though they were, in fact, close to bankruptcy. He also mined the libraries, the librarians reluctantly lending him unlendable books. When he came upon a book that he had to read but was not in circulation, he would sit down and begin to read it on the spot, a habit that sometimes got him the book. He cold-called editors regularly—professional writing can be a lot more like sales than people sometimes realize. "I find that I talk with these poor men as if I were over head and ears in business and a few thousands were no consideration to me—I almost reproach myself for bothering them thus to no purpose—but it is very valuable experience—and the best introduction I could have," he wrote to his mother. He even tried selling subscriptions to a magazine, the *New Agriculturalist,* which took him up to the rural suburbs of Westchester County as a door-to-door salesman. He was up for anything related to publishing, and he was ready to write almost anything. "The *Democratic Review* is poor, and can only afford half or quarter pay—which it *will* do—and they say there is a Lady's Companion that pays—but I will not write anything companionable," he wrote home.

He did try to write stories he thought salable. Ironically, when he managed to get himself into the city via ferry from Staten Island, he found resistance from Emerson, who, after all, had arranged for him to go in the first place; in his more and more perplexing manner, his mentor sent him a line about "forsaking the deep quiet of the Clove for the limbo of the false booksellers." Then again, Emerson himself had managed to forsake booksellers by turning to the lecture circuit, and up until now, he'd had Thoreau to help take care of his family. (Sometimes, while on the road, Emerson would write to his wife, Lidian, pining for the moment when he would return to his children and fall down on the floor to play with them, but then at home he would shut himself off, leaving family affairs to Lidian and the servants.) When Emerson gave him a hard time, Thoreau shot back defensively: "I was absent only one day and a night from the island, the family expecting me back immediately I was to earn a certain sum before winter, and thought it worth the while to try various experiments."

Though he sent out numerous stories, he sold very few. "My bait will not tempt the rats; they are too well fed," he wrote to his mother.

On the other hand, he met people, preceded by letters, on many occasions, from Emerson—letters that would often praise Thoreau's intelligence, while explaining that he was also a bit of a country bumpkin. Thoreau was thrilled to meet Henry James, the father of the novelist. The elder James was a Swedenborgian and Christian Socialist whose family had come from Ireland; a social reformer, he was intensely interested in the Brook Farm community, and the disciples of Charles Fourier, a French social theorist who proposed a reorganization of society based on cooperative labor communes. "It was a great pleasure to meet him," Thoreau said. "It makes humanity seem more erect and respectable. I was never more kindly and faithfully catechized. It made me respect myself more to be thought worthy of such wise questions."

He was not always thrilled to meet people involved in associations, but at the time just about everyone he met had a stake in a utopian community. Thoreau met these ideas with the circumspection of Emerson's side of the Transcendental divide; reform had to be based on the individual, he still felt. After he met William Henry Channing, a Transcendentalist and social reformer, for instance, Thoreau wrote to his sister: "They want faith, and mistake their private ail for an infected atmosphere . . .

"To speak or do any thing that shall concern mankind, one must speak and act as if well, or from that grain of health which he has left," Thoreau went on. This sounds theoretical, if not corny, until you remember that the man who is saying it is trying to make a living against all odds and at times is often having great difficulty breathing.

AN ASSOCIATIONIST HE ENJOYED associating with was Horace Greeley, the editor and political leader. They hit it off immediately. Their mutual respect was primed by all the positive notices that Greeley had given Thoreau's published work. Also, Thoreau would most likely have respected his Orestes Brownson–like push for social reforms and, for instance, the amelioration of the plight of workers (through workers' housing and unions). But mostly Thoreau loved that Greeley, this most prominent New Yorker, was a country boy, born in Vermont, raised as a printer in New Hampshire, running a Whig newspaper in the big city, influencing the nation. "Now, be neighborly," Greeley often said, which Thoreau appreciated. Greeley had always praised the *Dial*, and he would hire Margaret Fuller as his first female columnist, one of the first in the United States. He never seems to have offered Thoreau a job on his paper. But they remained friends for the next twenty years, so that Greeley even acted as Thoreau's literary agent when Thoreau eventually returned to

Concord, placing his stories in magazines and hunting down payments while taking no fee, a free-lance writer's dream. He referred to Thoreau as a "true poet."

Thoreau also met John O'Sullivan, the publisher of the *Democratic Review,* the preeminent literary and political magazine. Hawthorne had set up a meeting between the two men in Concord, just before Thoreau had moved to New York. O'Sullivan was a Jacksonian Democrat who is said to have coined the term *Manifest Destiny,* and when Thoreau stopped into his offices that fall, O'Sullivan was hoping for more nature writing along the lines of "The Natural History of Massachusetts"—the piece that he and Hawthorne so admired. (Emerson thought O'Sullivan was too crassly political—as Emerson would, of course—but Thoreau was flattered that the editor would want to print anything by him.) Thoreau submitted a book review, entitled "Paradise (to Be) Regained."

It was Thoreau's critique of the utopian ideals of John Adolphus Etzler and his book *The Paradise within the Reach of All Men, without Labor, by Powers of Nature and Machinery, an Address to All Intelligent Men.* Etzler was a technocratic utopian, and the book imagined a communitarian world wherein man's inventions and machines would fulfill society's needs by thoroughly utilizing nature. Thoreau's piece slams Etzler's ideas and to some extent those of Charles Fourier (who influenced Etzler), Thoreau rejecting a collectivism focused on securing material comfort. "The chief fault of the book is, that it aims to secure the greatest degree of gross comfort and pleasure merely," Thoreau said.

From Thoreau's standpoint, it all seemed to be a lot of engineering and effort for what might be achieved with less effort and little engineering. Thoreau's standpoint, please note, is funny: "There is a certain divine energy in every man, but sparingly employed as yet, which may be called the crank within,—the crank after all,—the prime mover in all machinery,—quite indispensable to all work. Would that we might get our hands on that handle!" He especially disagreed with the idea that one had to wait for reform, wait, in this case, for the accumulation of capital. "What is time but the stuff delay is made of?" he asks. *Paradise (to Be) Regained* is cutting and thoughtful, but what is most impressive to the free-lance writer today is that Thoreau had the chutzpah to send this rejection of Etzler to a journal that was likely to like him and his ideas. Initially, O'Sullivan rejected it. He encouraged Thoreau to make changes and asked Thoreau for something "literary," alluding again to his nature writing. Thoreau didn't make any changes to the Etzler piece, but somehow convinced O'Sullivan to take it anyway, which he did, running it as the lead.

In the meantime, Thoreau sent him another piece, a kind of writing known as the familiar essay, a popular form at the time, entitled, in this case, "The Landlord." It is a piece about a good landlord, in comparison to a bad one. Critics fall over themselves to talk about how bad "The Landlord" is and how it was merely a throwaway piece for Thoreau. He wrote it "to sell," he told his mother. In contrast, he didn't sell any of his ancient Greek translations in New York. "To sell" was not, however, entirely negative to the public Thoreau, to the Thoreau who saw the possibility of affecting a large audience—larger, for instance, than that of the dying *Dial.* He couldn't afford not to sell his pieces, as Emerson could; he had no backup lecture circuit.

It is possible to see, furthermore, that buried beneath the intentionally familiar and eminently more sellable style are the themes he discussed in the Etzler review. The landlord so pleasingly described is an antidote to the Fourierists' vision, in fact. In the Fourierists' future, technology fills all needs, and there is no use for the usefulness of man. (This is a gross generalization of the Fourierist view, but it is the aspect Thoreau was attacking.) For Thoreau the usefulness of man—the particular usefulness of each and every man—is what makes a place free and good. In a time when people are being put in factories, their skills compartmentalized and demeaned, when the tasks of men and women are being codified and coded, Thoreau is championing the peculiar and essential quality of the residents. "The traveler steps across the threshold and lo! he is master, for he only can be called proprietor here who behaves with most propriety in it." He is most himself who is most himself, a preview of the themes of *Walden*.

He wrote his sister to say that "The Landlord" wasn't worth the money to mail it to her, but had he mailed it, she would likely have recognized her brother's code. She would have noticed too the dig at Emerson, the landlord who had turned Thoreau into a servant. Travel was, as well, always one of Thoreau's favorite metaphors. Thoreau understood the value of true hospitality, having grown up in a town filled with travelers, with teamsters filling the taverns, in a home filled with boarders. The good landlord brings together "nations and individuals" who are "alike selfish and exclusive." The good landlord loves all equally. "Surely, he has solved some of the problems of life."

THOREAU'S MOST SIGNIFICANT MEETINGS while in New York were with the ocean and the masses. "There are two things that I hear, and am aware that I live in the neighborhood of—The roar of the sea—and the hum of the city," he wrote to Emerson on May 23. On Staten Island, he was on the coast of the Atlantic Ocean.

> Everything there is on a grand and generous scale—sea-weed and water, and sand; and even the dead fishes, horses and hogs have a rank luxuriant odor. Great shad nets spread to dry, crabs and horseshoes crawling over the sand—Clumsy boats, only for service, dancing like sea-fowl on the surf, and ships afar off going about their business.

It was a completely new landscape for him, and it was inspiring in the moment, and for what would later come: his book about the sea, *Cape Cod*.

Likewise, he saw the flow of humans that was arriving in New York City from the world, the tide of people who were looking for a kind of personal renewal but acting on less intellectual impulses: they were escaping poverty, and hoping or praying for a better life, or just staying alive. It's not clear why, but Thoreau thought the city was a better place to live than to visit. He found the appetite of the crowd appalling. He disdained the rush to spectacle; when he went to see a canoe race on the Hudson River between Chippewas and New Yorkers, he marveled that most people came to see the crowds. "Canoes and buffaloes are all lost, as is everything here, in the mob. It is only the people have come to see one another." He was social, but he wasn't *that* social; he was not a throng guy. In the city a crowd is self-perpetuating, an organism unto itself, like lichen, and this is difficult to recognize at first, even for a self-styled naturalist.

But a degree of disdain comes from the fact that he was there to write, to report, and so had that reporter's remove, compounded with his own homesickness. He would return to New York, after leaving in a few months, as a businessman, selling pencils for his family business, and as a tourist, off to see P. T. Barnum's American Museum, to attend an opera with John O'Sullivan, or even to walk Brooklyn with Walt Whitman. Instead of disparaging the crowd, he saw in it potential. The masses arriving in the city were the America-to-be. The English travelers were off to the fancy hotel, but most of the rest were living lives of desperation, living (literally) on the street:

> I have crossed the bay 20 or 30 times and have seen a great many immigrants going up to the city for the first time—Norwegians who carry their old fashioned farming tools to the west with them, and will buy nothing for fear of being cheated.—English operatives, known by their pale faces and stained hands, who will recover their birth-rights in a little cheap sun and wind,—English travellers on their way to the Astor House, to whom I have done the honors of the city.—Whole families cooking their dinners upon the pavement, all sun-burnt—so that you are in doubt of where the foreigner's face of flesh begins—their tidy clothes laid on and then tied to their swathed bodies which move about like a bandaged finger—caps set on the head, as if woven of the hair, which is still growing at the roots.—each and all busily cooking, stooping from time to time over the pot, and having something to drop into it, that so they may be entitled to take something out, forsooth. They look like respectable but straightened people, who may turn out to be counts when they get to Wisconsin—and will have their experience to relate to their children.

One of the biggest obstacles to Thoreau's success in New York was that he knew he could return home, which he did—at first temporarily, for Christmas. When he got back, Emerson, who was away, asked him to introduce Orestes Brownson at the Concord Lyceum. Thoreau was happy to do it. A few days later, he gave a lecture himself, on poetry, excerpts from which Horace Greeley published in the *New York Tribune*. Now he was not merely a *Dial* contributor but a minor star, and by then Emerson had written his brother to say that he could dispose of the few things Thoreau had left behind—Henry would not be coming back. Thoreau figured he was better off being a rising star while planted in his home universe. Thoreau moved back in with his parents. He had to help with the family business, which was struggling, and as a result he stopped writing for a while. Emerson was amazed: you can hear him throwing his hands up in the air. Emerson wrote in his journal that Thoreau was not a writer but a shoemaker.

When April came, Thoreau took a rowboat up the Sudbury River with his friend Edward Hoar. They caught some fish and built a fire in an old stump. They hoped to make a chowder, but what they made instead was one of the largest forest fires in the history of the town. They sprinted back to town for help, but on the way Thoreau stopped on a hill to look out at vast acreage of damage, realizing, no doubt, that he was too late. He could barely talk about it for years, and a lot of people in Concord never forgave him. For the rest of his life people would refer to him as "woods burner." Burning down the woods was a great way to ruin his reputation in his hometown forever.

Chapter 7

THE ROAD TO *WALDEN*

AT THE BEGINNING OF 1844, Thoreau was back in Concord, a pariah after burning so much of the town woods and an irritating puzzle to his mentor, who couldn't understand his penchant for nonliterary work, and he was on the cusp of success. If he left not sure where his visit to New York would take him, he came back to Concord knowing that it had ended up taking him home, his homesickness cured. A friend implored him to travel to Europe, but he said no, twice. He had long dreamed of living on a lake or a pond, and now he convinced Emerson to allow him to build a cabin on a piece of land Emerson had recently purchased, along the shore of Walden Pond—a wooded lot that cost Emerson $200, which he bought in part because the price of fuel was rising, owing to the lack of woodlots, and partly because he liked the idea of buying it. Walden was not yet *Walden,* and it's important to see it as Walden to understand where the trip to *Walden* began. Of course, Thoreau was a Transcendentalist and a budding popular writer of natural history, and, yes, he was steeped in the nature writing of the European Romantics, and he had ideas about writing of the pond and its environs, as a landscape painter would choose a scene for its proximity to what he or she considered good views. But he also went to the pond to work and to make a point about work—work in his own life, as well as in the lives of everyone else. It's important to think about the economic climate. As the country reeled from market forces, as the gap between rich and poor widened, as people strained to make a living and (whether they were successful or not) saw their social and family life begin to change as a result, Thoreau was about to give a very practical answer to the question that Emerson asked, the question that was not just on the mind of philosophers past and present but on the mind of the country: "How shall I live?"

Staten Island had exhausted him. Aside from going there ill, he had worked as a tutor while simultaneously trying to meet editors in New York City and publish articles. Meanwhile, he was writing pieces for Emerson and the dying *Dial,* even if he may have wanted them published more prominently or in more lucrative venues. At least they were getting published. As a species, writers often like to concentrate their energy, to focus, and Thoreau must have felt as if he were striving vainly again, at the end of his time in New York. Now, though, he was focused on what he could do, what he *would* do, if he could find a time or a place. He had begun working on a book while on Staten Island, an account of his trip up the Merrimack River with his late brother, John. On his return to Concord, he had begun collecting notes on the trip, as well as collating his journal entries. At Walden Pond, he would settle into work. No more working as an *au pair* to Emerson's children, which he enjoyed as much as he found it distracting. No more living at home in the family boarding-house, where he was expected to talk to guests, where the family always ate together, where there was always a social or a meeting coming up. No more staying in what he ultimately considered a dull household—that is, Emerson's brother's house. In his journal, he lamented of Mr. and Mrs. William Emerson's house: "O I have seen such a hollow glazed life as on a painted floor which some couples lead—with their basement parlor with folding chairs—a few visitors' cards and the latest annual . . . There they do not live. It is there they reside."

Thoreau wanted to try to live in a place, rather than reside. He came back from New York with the idea that killing yourself for your career might not be the only way—living in New York for a while will do that to you. He was reformulating his work life, and the cabin at Walden Pond seemed to be a possible, if temporary, solution. In itself, work was his main reason for the moving to Walden Pond—to work to his heart's writerly content, something his friend Ellery Channing had been egging him on to do. Channing was from a prestigious Boston family, and much was expected of him as far as work went. He aspired to be a poet, though he had neither Thoreau's talent nor his determination. Despite initially suggesting to Emerson that Thoreau was limited as an artist because he was not married, he became Thoreau's best friend, his almost daily walking companion in the woods and on occasional hiking trips. Thoreau, meanwhile, helped Channing through a bad marriage, which is ironic, of course.

It was Channing who suggested Thoreau go to Walden and "begin the process of devouring yourself alive."

To GET AN IDEA OF WHAT THOREAU was thinking about work and about making a living when he wrote *Walden,* you have to stop and look at the work situation in Concord and the towns in the area, to imagine the economic landscape around the pond the way you might try and imagine the trees and the birds and the water. To imagine Thoreau and his writing without considering the economy is a little like thinking about *The Grapes of Wrath* without considering the Great Depression. The Concord that existed when he went to college, in 1834, was different from the Concord in which he is about to build a cabin, in 1845, at twenty-seven; like the rest of America, it is in the midst of a transformation. It bears repeating that from 1837 through 1843, the country was caught in a severe financial depression. In general New England's economy was changing, colonial agriculture being replaced by the early stages of modern industrial capitalism, all the economic and political power that had been dispersed among farmers now being concentrated in a smaller number of people, primarily landowners and business owners. The revamping of the United States in the time between the Revolution and the Civil War has been called "a total ecosocial transformation." The *eco* here is as in *economy,* not *ecology,* a word one might be more likely to suspect to show up in a book about Thoreau. Then again, a little rooting around in the etymology shows they share origins: the Greek word *oikos,* which means house.

Transportation facilitated the change in New England. Port towns were connected with more and more toll roads throughout the 1700s, and then, in 1825, New York State opened the 366-mile-long Erie Canal, which consequently opened up western New York State for farming; in one moment, newly planted wheat fields in the formerly faraway western New York State were in competition with the New England land that had been farmed for generations. The New England farms suffered; work went west as flour was shipped east, and another migration of former farmers went west when the railroads came in the 1850s. Changes in work life changed lives: boys grew up on farms and worked on them or, in larger towns like Concord, went from apprentice to clerk, rising after their apprenticeship, moving from father's place to merchant's home. By the 1850s, boys lived at home longer, like Thoreau, and when they got jobs, as clerks or, more likely in the cities, at temporary work, they were not members of merchants' households, but lived alone in boarding hotels and boardinghouses, eating in restaurants, drinking in saloons—living lives that seemed to their elders simultaneously exciting and frightening, a break from the old ways. Women too lived at home longer, single men having gone west or to the city; they were more likely to be spinsters, like Sophia, Thoreau's sister. If women did not work in factories, or as garment workers for hire, they went off to teach, as Thoreau's sister Helen did, until sickness caused her to come home to tutor until she died. Women also became domestic servants, in the new bourgeois households of the cities—in the homes of the factory owners and managers or of the lawyers, doctors, and schoolmasters who were finding work in the more and more populous cities, which led to Emerson's aforementioned "servant problem."

The idea of how people made anything was about to change, the concept of craft itself was suddenly transforming. Prior to Thoreau's time, young people had been raised in an artisan culture; the years between 1740 and 1830 have been referred to as a golden age for American craftspeople, with production based in family workshops. Apprentices lived with the families of their mentors; age and learned skill were valued alongside cooperation. But as the Civil War approached, Jefferson's yeoman farmer and the artisan went from being the respected citizens of young America to the outsourced workers. Crafts were dying and farms were shutting down, and young people were suddenly without their old paths to employment. Due to the change in the nature of work, there was a change in the nature of cities. As Thoreau had noticed while living in New York, the city was crowds, crowds of workers and crowds of immigrants (whom many people did not, or chose not to, notice), crowds of people who were mostly moving through; urban populations in the United States increased by sixty percent in the 1830s, by ninety percent in the 1840s. Cities became transient places, and the idea of working-class migrant communities living together in neighborhoods is mostly myth. The city also represented a magnified version of what was going on in Thoreau's hometown.

As Thoreau would have noticed just walking down the street in Boston, or even in Concord, unemployment was suddenly a way of life. In 1837, because of the financial crisis, nine-tenths of the factories in the eastern United States were closed. Ninety percent of the factories in New England closed. The mills were all but completely shut down in Boston and Lowell, and two thousand were unemployed at Lynn, wages in the area reduced by half. The work that people could get was not necessarily worth it. In 1830, one contemporary commentator estimated there were twenty thousand working women in Philadelphia, Boston, and Baltimore, their average wage being $1.25 a week—allowing for sickness, unemployment, or care for children, lodging at 50 cents, fuel and clothes, she would have 2¾ cents per day for food. In 1845, as Thoreau returned to Concord from Staten Island, conditions were worse, and the *New York Sentinel* reported that of the fifty thousand workingwomen just in New York, half earned less than $2 a week. It wasn't until the 1870s that wages for women increased to $3 a week. The social effects of the depression that started with the Panic of 1837 and lasted until after Thoreau lived at Walden are said to have exceeded the Great Depression, and people considered it proof that capitalism might not be working. "The nation has been drawing on the Future, and the Future dishonors the draft," said a contemporary commentator.

As this depression hit the new middle class, Horace Greeley noted that one-fourth of all mercantile and manufacturing businesses were bankrupt. " 'Hard Times!' is the cry from Madawaska to Galena," he wrote. By the time he met Thoreau, Greeley was calling employment the one essential question—"of more importance than any ruling political topic." "Fly," Greeley wrote to his readers, "scatter through the country, go to the Great West, anything rather than remain here." He may be known for saying "Go West, young man," but here he was saying, essentially, "Get the hell out of here," as in "Run for your life!" Politicians seemed to be just talking, or, in the case of the Whigs, not talking. Jacksonian Democrats ran Martin Van Buren, and the elite New Englanders' Whigs fought pseudopopulist fire with pseudopopulist fire, throwing barbecues and clambakes and serving cider at log cabins built in town squares in honor of their war-hero candidate, William Henry Harrison, who was instructed not to speak about any issues.

A wave of people took Greeley's advice, including religious and other groups. (Others were advising that people head west too, such as real estate speculators.) Just as the Transcendentalists responded to the deteriorating social and economic situation with their utopian ideals and experimental farms, so Joseph Smith, the founder of Mormonism whose family had moved to western New York State from Vermont, led his followers from New York State into the deserts of the West, his transcendental utopia. In the same way, the Oregon Trail can be viewed as a pipeline of American nationalism, or as a long unemployment line, a trail of economic desperation and hope (Native Americans sometimes shared food with starving settlers-to-be). In 1838, the Society for the Prevention of Pauperism in Boston set up committees to help people find work or leave the city, and by the mid-1840s, the idea of permanent unemployment was beginning to settle in. In New York, in the 1840s, commentators noted that there always seemed to be at least twenty thousand people out of work, and three hundred thousand others living on less than a dollar a week.

VIOLENT CLASS WARFARE was more of a possibility than the typically genteel study of the Transcendentalists' time would indicate, or marketers who invoke Thoreau's name nowadays might imagine. Union membership had taken hold on a mass scale—in 1834, New York City's General Trades Union created a National Trades Union, and had a march a mile and a half long—but the layoffs zapped their power. In the winter of 1837, as theaters were deserted and markets empty, renters were planning a mass action in New York City; landlords collectively held back on their attempts to collect. For the first time in U.S. history a president,

Andrew Jackson, used federal troops in a labor dispute—against the immigrant Irish workers on the Chesapeake and Ohio Canal who had attacked scab workers. The Workingman's Party was agitating in Boston, while women working in the mills in Lowell went on strike, what they then called a "turn out," singing, "Oh! isn't it a pity, such a pretty girl as I/ Should be sent to the factory to pine away and die?" Philip Hone, the former New York City mayor and diarist, wondered how a workingman fed his family, given that Hone's upper-class friends were in dire straits. "What is to become of the working classes?" he wrote. That spring, Emerson wrote in his journal:

> Cold April; hard times; men breaking who ought not to break; banks bullied into the bolstering of desperate speculators; all the newspapers a chorus of owls. . . . Loud cracks in the social edifice.—Sixty thousand laborers, say rumor, to be presently thrown out of work, and these make a formidable mob to break open banks and rob the rich and brave the domestic government.

Back in the spring of 1837, Orestes Brownson (just after Thoreau had finished working for him) had even gone so far as to call for a revolution—or he was perceived to have called for revolution, at least. He cited what he called a general evil in the system, and thought that the constituency with the economic resources to change matters was more interested in the status quo. As is the case today, the vested interested were likely to stay vested. Those with economic resources and opportunities and the mass without were, he said, "waiting but the signal to rush to the terrible encounter, if indeed the battle has not already begun." He went on:

"If a general war should now break out, it will involve all quarters of the globe, and it will be in the end more than a war between nations. It will resolve itself into a social war, a war between . . . the people and their masters. It will be terrible war!"

That Brownson mentioned war, which translated to everyone listening as *revolution,* caused the likes of Emerson to step away from him, and Boston intellectual society stepped away too. Very quickly, Brownson was exiled within the Transcendentalists—they were radical, but they weren't *that* radical. The radical who replaced him was George Ripley, who was about to found Brook Farm. Brownson was derided as someone who wanted to sink the ship of society, but Ripley was perceived as someone who wanted to raise it up. Even if they didn't go to Brook Farm to partake in the social experiment, they preferred Brook Farm to Brownson, who was saying nice things about the French Revolution. On the other hand, a lot of people went to Brook Farm, to live or to take a peek at how it was going.

THOREAU AND RIPLEY WERE FRIENDS, and although Thoreau's refusal to join the Brook Farm community is, as stated earlier, often viewed as proof of the antisocial Thoreau, a recently discovered letter has shown that Ripley considered Thoreau a supporter. In fact, Thoreau visited the farm just before leaving for Staten Island. He was ill at the time—so ill that Ripley and the Brook Farm residents were worried about him making it home safely. As opposed to Brownson, who saw the problems in society as a class struggle, where the economic structures pitted a small ownership class against a growing laboring one, Ripley, who was just as distressed by the national economic situation as

Brownson, felt society's problems were rooted in the way people thought about work. People with enough money to make a choice about whether to do labor or not began to think that manual labor was beneath them. This had not been the case, for example, for the yeoman farmer of Jefferson's day, and Jefferson's day was still in recent memory. (Jefferson died on July 4, 1826.) More and more, wealthy Americans perceived themselves as intended only for intellectual pursuits, for work with their heads as opposed to their hands. To Ripley, the sharp judgments—on both sides of the widening social fissure—made for the sharp divisions in society. He hoped to shake up American society by offering another way, another kind of structure for daily life—an ideological program of improvement, based on the equitable distribution of work.

"God has given each man a back to be clothed, a mouth to be filled, and a pair of hands to work with," Theodore Parker wrote. Thus, at Brook Farm, a Harvard-schooled minister like Ripley would regularly plow fields and shovel manure. He would build a fence and milk the cows. It was a little like a dude ranch, but with women too, and for life, or at least that was the plan.

That's why Ripley worked so hard to convince Emerson to live at the farm. Emerson was the nation's best-known intellectual, a celebrity, and to have him would have meant a public relations coup for the healing of the divide between people who worked and people who paid people to do it. Ripley wanted Thoreau for the farm, on the other hand, because of his poetic abilities and obvious intellectual strength, but also because Thoreau was someone who actually knew how to work, who had aspired to own and run his farm, who knew a craft (pencil making), who knew how to fix and build machines, who could survey land, who had already shoveled a lot of manure (as opposed to Brook Farmer Nathaniel Hawthorne, for instance, who was about to learn how). By virtue of his upbringing, Thoreau was a poor man's rich man.

Ripley knew that Emerson, for all *his* talk about the benefits of work, was work-shy, as well as uncertain about mixing with the manual laborers. It was because of the attitude of people such as Emerson that the Brook Farm community charged a $500 entrance fee, a stock investment, restricting many manual laborers. It probably restricted Thoreau as well; he would have not been able to afford it. Ripley mentioned the fee to Emerson, spinning the situation hard:

> I believe in the divinity of labor; I wish to "harvest my flesh and blood from the land"; but to do this, I must either be insulated and work to disadvantage, or avail myself of the service of hirelings, who are not of my order, and whom I can scarce make friends . . . I wish to see a society of educated friends, working, thinking, and living together, with no strife, except that of each to contribute the most to the benefit of all.

Ripley was being arrogant. On the other hand, he was attempting to cajole the arrogant, and to Ripley's credit, a more egalitarian appeal comes in the postscript, when he explains to Emerson why there will, in fact, be people on the farm who are not intellectuals, who are not of Emerson's "order," who are what Emerson might have called "mere workers":

> I think we should be content to join with others, with whom our personal sympathy is not strong, but whose general

> ideas coincide with ours, and whose gifts and abilities would make their services important. For instance, I should like to have a good washerwoman in my parish admitted into the plot. She is certainly not a Minerva or a Venus; but we might educate her two children to wisdom and varied accomplishments, who otherwise will be doomed to drudge through life.

Emerson, of course, declined, hoping to start his own miniature social experiment, even if he couldn't convince his cook to eat with him. Emerson considered the "servant problem," as it was now, of tremendous significance. He made a list of the important issues in America in 1838: "War, Slavery, Alcohol, animal food, Domestic hired service, Colleges, Creeds, & now and at last Money." But aside from the queasiness that manual labor no doubt gave him, Emerson, understandably, probably did not see the farm solution as the answer for a married man with children and extended family living at home.

By the time Thoreau got to Walden, Emerson had given up his experiment with "family-style" domestic service and returned to the pay-them-a-wage style. Emerson knew the argument *for* domestic servants, especially the argument for domestic servants in the service of people such as writers and ministers. In an advice manual of the time, a servant is consoled by the reminder that her position is not "a low and mean thing to be." If there were no servants, the manual argues, the minister and the writer would have no time to preach and write. "Yes," a servant in the manual says, "I see that servants help to get the gospel truth taught." Emerson also saw that waged servants would inevitably have their wages undercut—before there was even a railroad to build in Concord, the area of Walden Pond was flooded with Irish laborers. His friend Edmund Hosmer was threatening to leave town, his labor always being undercut by the Irish laborers, who were themselves treated, they noted, as slaves. Like running water, people paying people—without the benefit of restriction or penalty or coercion or compassion—seek the lowest pecuniary ground.

Thoreau, for his part, visited Brook Farm and hung out with the residents, with whom he felt a kinship, intellectual and otherwise. (If they were to have imagined an ideal Brook Farmer, he might have been it, except for being unable to afford the entrance fee.) In "Paradise (to Be) Regained," his critique of Etzler, Thoreau had praised manual labor and criticized theorists who were not willing to work. The Brook Farmers—or some of them, anyway—were workers. In his journal, Thoreau was more critical of abolitionists with domestic workers than of the likes of Ripley. (Emerson's disdain for abolitionists and his mixed feelings about domestic labor contributed to his reluctance to appear in public as an abolitionist.) "Two tables in every house!" Emerson wrote in 1844. "Abolitionists at one & *servants* at the other!" In December 1843, just after Thoreau had returned home from living in New York and after having paid a visit to the farm, farm resident George Bradford wrote to Emerson, worried about having sent a sick man home in a snowstorm—it speaks to how frail Thoreau could often be and how close he was to the reformers. "We are quite indebted to Henry for his brave defense of his thought which gained him much favor in the eyes of some of the friends here who are of the like faith," Bradford said.

When he built his house on the pond, Thoreau would work happily in his bean field, as the people passed by in gigs—gigs being the nicer, more expensive horse carriages, the luxury sedans of their day. He joked that they passed by watching him as if, he says, he were a strange creature, a laborious being, a transplanted Brook Farmer: an "*agricola laboriosus.*" Or, as he would eventually write in *Walden:*

> But labor of the hands, even when pursued to the verge of drudgery, is perhaps never the worst form of idleness. It has a constant and imperishable moral, and to the scholar it yields a classic result. A very *agricola laboriosus* was I to travellers bound westward through Lincoln and Wayland to nobody knows where; they sitting at their ease in gigs, with elbows on knees, and reins loosely hanging in festoons; I the home-staying, laborious native of the soil.

FOR A LONG TIME BROOK FARM WAS VIEWED as a well-intended debacle, a high-minded failure, a "theoretical brotherhood" that, practically speaking, split, leaving behind disgruntled siblings who poked fun at it, as Hawthorne did in the *Blithedale Romance.* "Our labor symbolized nothing, and left us mentally sluggish in the dusk of the evening. Intellectual activity is incompatible with any large amount of bodily exercise. The yeoman and the scholar—the yeoman and the man of finest moral culture, though not the man of sturdiest sense and integrity—are two distinct individuals, and can never be melted or welded into one substance." The name Blithedale was itself a joke, meaning happy valley. Hawthorne had gone to Brook Farm to save money for his upcoming marriage, among other reasons, and in Hawthorne's book, Blithedale falls apart due to members' self-interests. Just as Thoreau is a hermit, the odd duck at the pond, the thinking has long gone, the Brook Farmers were quacks. But Brook Farm was not the mere fad that Hawthorne readers might imagine. It started out as a loose association of idealists, but after it converted to Fourierism, in 1844, Brook Farm went from a retreat for a few intellectuals to what has recently been called "a burgeoning center of agricultural and industrial production." Mostly, the problem was that the classes don't really want to mingle, then or now. But the underplayed success of Brook Farm and the extent of Thoreau's cheerful engagement with the community, heretofore unknown, raises the question: What if we thought of the utopian experiments in the time before the Civil War as successful to some extent, rather than merely thinking of them as failed? Utopia is by definition never to be achieved, but what is to be achieved on the road to utopia?

There were thirty such communities, called phalanxes, around the country, populated by as many as one hundred thousand people, the trend being driven in large part by all the frustration brought on by bad economic times. They ran across the northeast to Iowa and Wisconsin and to Texas, where La Reunion, in Dallas County, counted 2,240 members, many of them French.* The Brook Farmers ran a magazine at their

* Fourierism is one of those movements that suffers from being related to a lot of movements, each of which disparages it in retrospect. Marxist historians have said it was just an impractically idealistic forerunner of what they consider the more scientific socialism of Karl Marx and Friedrich Engels. Labor historians say it was too middle class to be significant, which is comical (to me) in that in the twentieth century the labor movement helped build America's middle class. At the time, the Fourierists were worried about seeming too French, so they tried to show that Fourier's ideas matched up well with the American dream. (They deemphasized Fourier's interest in numerology and free love, an aspect the prudish Emerson critiqued as "a calculation how to secure the greatest amount of kissing that the infirmity of the human condition admitted.") On the other hand, in *The Utopian Alternative: Fourierism in Nineteenth-Century America,* Carl Guarneri calls it "a coherent radical critique of American social organization." Guarneri said it was popular because it addressed issues that the

farm in West Roxbury, Massachusetts, the *Phalanx*. Members of the farm went out lecturing. The schools at the farm were profitable and reputable: Orestes Brownson sent one of his children to school there. Regular coach service brought visitors from Boston; the farm made money shuttling in curious political thinkers and friends and relatives of the working-class members who took part after 1844. It was a tourist destination in the manner of a fancy organic food co-op today or an organic farm and associated organic restaurant, or a Shaker farm back then. The farm associations were a topic of the day, and everyone who considered themselves well-read and public-minded took a position on them. People on the *for* side included the likes of Elizabeth Peabody, who pioneered kindergarten in America. She believed that if everyone kicked in on work, then another kind of wealth would surface. "The hours redeemed from labor by community, will not be reapplied to the acquisition of wealth, but to the production of intellectual goods," she wrote. "This community aims to be rich, not in the metallic representative of wealth, but in the wealth itself, which money should represent; namely, THE LEISURE TO LIVE IN ALL THE FACULTIES OF THE SOUL."

The movement got lost in the ensuing national debate on slavery and the resulting Civil War. Fourierists were against slavery, but they were against the wage system in the north as well, and Southern slavery apologists discredited the Fouri-

---

average American found troubling, like recession, unemployment, and how to make a living in the face of vast accumulations of wealth in the hands of a few. Get-rich schemers were also attracted to the movement. Remnants of the movement are in the municipal parks movement of the Gilded Age, in addition to feminism and workers' cooperatives and anarchism. One lasting tenet of Fourierism that we take for granted (even if we don't always practice it) is that labor needs as a balance the enjoyment of artistic pleasures, intellectual stimulation, and general cultural activities.

erists as a result, helping a faltering experiment falter further. But at the time that Thoreau went to Walden, the newspapers and journals were filled with debates and accounts and news of the Associationist movement. Horace Brisbane, the editor of the *Phalanx,* wrote a front-page column in Horace Greeley's *New York Tribune,* proclaiming the benefits of "Attractive Industry," "Democracy of Association," and the "Equilibrium of Passions." (Greeley himself had a colony, called Sylvania, composed of three hundred people, and he probably wanted Thoreau there too, though it was abandoned after frost killed all the crops.) People talked about associations the way they talk about green living today. And the farms actively addressed the intellectual—and practical and economic—battles over slavery, among other issues of the day, including the role of women in the workplace, education reform, and labor and class issues. In some ways Ripley did not believe the farm itself was the answer; he hoped Brook Farm would be the beginning of something. To that extent it was successful, at least until abolitionism, with the onset of the Civil War, became an issue in America that overshadowed all others. "If wisely executed, it will be a light over this country and this age," he wrote of the farm. "If not the sunrise, it will be the morning star."

What this all means in terms of Thoreau is that his house at Walden was not separate from the Associationist movement. It was a complement to it. The house at Walden was his version of the great social experiment of the day, his experiment with material poverty as well as spiritual and intellectual poverty, the one-man retreat that would lead to a new kind of society, a one-man farm. It was undertaken not in hopes of finding a way to be less social but in the search for a way to be *more* social, to find a way to real community, especially for the middle class who were, as Thoreau saw it, becoming obsessed with slaving away at one thing—in part by hiring people at cheaper prices to do work that only a generation before they might have done themselves, everything from growing food to cleaning their kitchen. When Thoreau went to Walden Pond, the economic divisions that the depression had exacerbated were still widening: the poor were struggling and wealth was concentrating in fewer hands. The middle class was striving for more but coming up, by Thoreau's estimate, with less. "The mass of men live lives of quiet desperation," he would soon write.

IN NEW YORK, THOREAU HAD FAILED IN HIS ATTEMPT to become a commercially successful writer, and now he set about being what he would end up living out his life as: a part-time writer. More exactly, he was to be a part-time writer and a part-time laborer, a free-lance, a term that in his time denoted mercenary.

A job that he would always have on and off throughout his life was pencil maker. Thoreau was an excellent pencil maker. He worked with his father, and added significantly to the quality of the family product. He was not just one of the excellent pencil makers in America; John Thoreau and Company pencils were the best-known pencils in the United States, praised by artists and artisans. Though he is remembered as a technophobe who shunned machines, Thoreau was intimate with the pencil-manufacturing business, and he knew it from every angle; he had gone on sales trips with his father (to New York, for example), and he had worked on the packaging and shipping at home. The pencils were bundled with paper wrappers advertising John Thoreau and Company. Most significantly, Thoreau used his engineering skills to develop a model that became known as the best pencil in the country.

He had begun working with his father after returning from college—it is thought that he investigated European pencil-making techniques in the library at Harvard. While first living at the Emersons', he would leave from time to time to work at his parents' house making pencils. On his job hunt in Maine, he noticed the local graphite implements: "I observed here pencils which are made in a bungling way by grooving a round piece of cedar then putting in the lead and filling up the cavity with a strip of wood." When he lived in New York, he kept an eye open for any interesting pencils, writing home and asking about "improvements in the pencil line." In 1849, German pencil makers began to ship their pencils to America, and the Thoreaus would abandon the pencil business and sell the graphite alone—a new kind of printing, electrotyping, made it feasible for publishers to put out cheap mass editions of books without laboriously resetting type, and the Thoreaus segued into this new market. When his father died, in 1859, Thoreau would take over the business, with his sister and mother. Around that time, Thoreau took out a subscription to a magazine called *Businessman's Assistant*.

When Emerson got back from New York, he said that making pencils seemed to be all Thoreau could think about, though he almost never talks about pencil engineering in his journal. It was a kind of left brain–right brain separation of church and state, or of art and pencils. On the pencil side, he developed a way in which to inject lead into the wood of pencils, the seamless pencil being a goal of pencil makers at the time. The pencil that was popular prior to Thoreau's development was manufactured by cutting the wood in half, filling it with graphite and gluing the wood together again. Thoreau studied various graphite hardnesses and invented a machine that manufactured a finer grind. The centerpiece was a cylinder, in which the finer graphite settled to the bottom for ready collection. The new invention pushed the company ahead of its rivals. The Thoreau pencil was more expensive—an award-winning brand, a practical luxury item, in a sense—and it rivaled European pencils.

BUT FIRST, A VACATION, with an eye, as always, to reporting. After returning from New York that winter and burning down a big swath of Concord woods in the spring, he set out on a summer trip with Ellery Channing, and by train, boat, and foot, explored the Berkshires and the Catskills. They visited Katerskill Falls, the scenic Hudson River painters' viewpoint that was being logged up at the time, the mill kept out of the painters' pictures—many naturalists at the time liked to make nature look even better, as if it needed help. Thoreau stayed with Ira Scribner, the sawmiller, and admired Scribner's rustic cabin, as well as a woman he met. He would describe her later in *A Week on the Concord and Merrimack Rivers,* his first book—plugging it in as if it had happened while he was on the Concord and Merrimack Rivers. "Its mistress was a frank and hospitable young woman, who stood before me in a dishabille, busily and unconcernedly combing her long black hair while she talked, giving her head the necessary toss with each sweep of the comb, with lively, sparkling eyes, and full of interest in that lower world from which I had come," Thoreau wrote. People on the boat Thoreau and Channing took on the Hudson River mistook Thoreau for a deckhand.

Back in Concord, he prepared to write *A Week,* the core

of his post–New York writing plan. Also, he did odd jobs—manual labor, mostly. As always, he was around. Thoreau rang the bells for an address by Emerson, who had by now bought the parcel of land at Walden. He helped his father build a house on newly developed land behind the train station—his father took out a mortgage. After years of moving from rental house to rental house, the Thoreaus were about to have a home of their own, albeit with a sizable mortgage. The Thoreaus had survived the economic crisis in the manner of the economic day—by specializing, and thriving, in the manufacture and sale of pencils, thanks in large part to their remaining son.

Thoreau worked more hours in the summer of 1844 to help come up with the down payment for the house the Thoreau family was building. In September, when his father purchased a small lot, and signed a $500 mortgage, Thoreau dug and stoned the cellar of the new house, and with his father built the two-story square building that biographer Walter Harding described as "quite lacking in distinction of any sort." The Fitchburg Railway Company was auctioning off the shanties of Irish laborers at the time. The Thoreaus bought a few of them and constructed a shed in which they ran the family pencil business. Thoreau planted apple trees around the house, a finishing touch, all the while helping churn out pencils. "Thus," writes Henry Petroski, in *The Pencil: A History of Design and Circumstance,* "contrary to the conventional wisdom then and still current around Concord and elsewhere, Henry David Thoreau was no slouch, even though in May 1845 he left home and the pencil business and began to build his cabin near Walden Pond, where he would live until 1847." I can't help wondering if Thoreau noticed that the word *mortgage* comes from the old French *dead pledge.*

He was an engineer who relished technological innovation, and his family was part of the market drive that Brownson and his supporters critiqued. He was not a hermit who had no experience in business. As opposed to farmers who had difficulty adapting to changing markets, or the local craftsmen whose work was replaced by less expensive versions of the same crafts brought in by train or on the canal, the Thoreaus had been able to keep their business alive—they were the ones putting someone *else* out of business. Yes, part of his "experiment" at Walden was a critique of the business world, just as he was about to step out of it. He would step away as a writer, but he would step away with practical work in his personal portfolio of labors. Emerson saw that he was not like other businesspeople, not like other writers. "Satan has no bribe for him," he said.

SOMETHING MISSING IN THE USUAL PORTRAIT OF Thoreau is that he was as practical as he was philosophical, as silly as he was serious. From his vantage point as a part-time free-lance writer looking for a new project—an "experiment in living," as he was to call it—a place at the pond seemed like a natural choice. Walden was one of his lifelong hangouts. It was close in but at a distance. It was a practical route to philosophical points. It was a stunt, plainly put, even a humorous stunt, albeit a stunt with meaning, and with a nice view. This was a faraway wilderness retreat right on the edge of town. "Why not live a hard and emphatic life," he wrote in his journal, "not to be avoided, full of adventures and work, learn much in it, travel much, though it only be in these woods?"

A literary stunt is a thing that happens all the time today in publishing circles: a writer living in a particular way—or partaking in a particular community or ritual or what have you—in order to ultimately report on the event or place or people. It is an essentially artificial experiment undertaken with an interest in making money on publication or putting forth a not-so-artificial argument (optional) or, in some cases, both. (On TV in the twenty-first century, we refer to one version of it as *reality*.) The genius of Thoreau's pond-side vantage point—a self-made house on a woodlot at the end of town—is that it affords so many possible critiques, not the least of which was aimed at the economic order he was a part of. He could discuss all this from the distance of his cabin, and with a naturalist's gloss, a sheen of faraway and extravagant life, when he wasn't actually far away at all. So much about the house at the pond was itself, before any writing began, a critique. All the issues of the day, everything that the Associationists and the Fourierists were debating about, are addressed in first, building, and second, residing in the cabin. In living at the cabin and writing about it, even for a short two-year period, he was rejecting the changes that nineteenth-century America presented to him. At the cabin, Thoreau rejected the outsourcing of work, and the degradation of work in general: the loss of craft, the demise of volunteerism. He rejected the creation of association after association in favor of real association with neighbors and friends, and, thus, with life and the world. The world was being made into a video game of the world, as he saw it, and he was going off to play the real thing. He would get some work done and report back.

His career choice then became a choice of not choosing. He still hoped to be a man of letters, and later he would get an idea he could make a living lecturing, but as of now, in this short term, he was choosing a career in no career. His famous letter, responding to Harvard's request to know his occupation, is as frivolous sounding as it is a secretly serious comment on the specialization that had hit antebellum America, that characterizes America today when parents council their high school kids on course choices that are suitable for their imagined future careers. Just as the very idea of a career is being born—careers not just for ministers and doctors and professors but for everyone—Thoreau rejects it, and in a way laughs it off:

> I am a Schoolmaster—a Private Tutor, a Surveyor—a Gardener, a Farmer—a Painter, I mean a House Painter, a Carpenter, a Mason, a Day-Laborer, a Pencil-Maker, a Glass-paper Maker, a Writer, and sometimes a Poetaster.

He would return from Walden after two years, but he would strive never to return to a full-time occupation. It riled people in town, as it does today. People are never more expert than when discussing how others are running their lives, and a noncareer choice, especially if a person can pull it off, is the kind of thing that confuses people and sometimes even makes them mad, especially if it contrasts with their own life choices. The old-timers in Concord seemed to get it, as, like Thoreau, they considered work an end in itself, which was why Thoreau often punned on his name as he pronounced it: "I do a thorough job."

He began building his house in March, a few months before he actually moved to the pond. He called it a house. His family called it a hut. Emerson called it a hut. Alcott called it a hermitage. Ellery Channing called it a "wooden inkstand," on account of how much work Thoreau got done there. Thoreau referred to it variously as a lodge, a hut, an apartment, and a dwelling, but more often than not to him it was a house. It was not a log cabin; he had not yet seen and admired the log cabins he would eventually come across in Maine. It was a small house that was built in the architectural fashion of the day: as a miniature country house. The tendency today is to think of Thoreau's lodging as an odd outpost, a crazy man's crude construction. But the cabin itself is the first authorial act of *Walden,* and it is, like the book, organized, thought out, streamlined, while philosophically weighty. In his first published writings, he had managed to give a subversive twist to nature writing, and now he was about to twist another kind of writing that was equally popular. Just before the Civil War, people were writing about their houses and their gardens in books and newspapers. Home wasn't just a place to live your life anymore; it was becoming a place to indicate your life's *style.* In building a house at Walden, he was, with his tongue in cheek, saying precisely who he was.

IF THE SO-CALLED MCMANSION, the exorbitantly-roomed home that was born in America's most recent expansionist economy, was the signature home of the 1990s—home to Internet billionaires and corporate CEOs alike—then the home of the pre–Civil War American who was of similar standing in his community (that is, on the top) was the cottage, which was sometimes a small cottage and sometimes verging on McMansion-size. The thinking of Thoreau's time, as is the case in home magazines (and home Web sites and home TV shows) today, was that the house made the man. In 1845, the woman generally made the house, by choosing decorations and designs, unless it was a house for *only* the man, even the married man, who would use a little house to "get away." In either case, the house mattered to people, especially those with the means to make it matter. "Uncouth, mean, ragged, dirty houses, constituting the body of any town, will regularly be accompanied by coarse, groveling manners," wrote Yale president and theologian Timothy Dwight. "The dress, the furniture, the mode of living, and the manners will all correspond with the appearance of the buildings, and will universally be, in every such case, of a vulgar and debased nature." In his essay, "The Moral Influences of Good Housing," Andrew Jackson Downing, a well-known landscape planner at the time, wrote: "We believe in the bettering influence of beautiful cottages and country houses—in the improvement of human nature necessarily resulting to all classes from the possession of lovely gardens and fruitful orchards." Downing felt you could discern the internal goodness of a man from the external signs. And while Downing called for simplicity, he also called for external signs of domesticity, such as entwined vines and gabled rafters, simple details that, one might argue, became complex in their pursuit of purity.

Printing was, to reiterate, in full swing and gaining, and helped to promote the ideas of what was becoming a growing consumerhood. Newspaper, magazines, inexpensive books, and pamphlets were everywhere in antebellum America. Thoreau read them all, including periodic literature that showcased the progressive models for domestic architecture—the real estate sections and "shelter magazines" of the time. He was like the cable news junkie who disdains political posturing and watches all the Sunday morning politics talk shows. Likewise, pattern books, books with plans for these houses, with architectural drawings for houses of all sizes, were being published in England and the United States. The average reader knew of the "cottage," "villa," or "country house," which were, according to Downing, especially numerous in Thoreau's environs, the quickly suburbanizing area around Boston. "For that species of suburban cottage or villa residence which is most frequently within the reach of persons of moderate fortunes, the environs of Boston afford the finest examples in the Union," wrote Downing in 1848.

It is to his credit that Thoreau was in no way revolutionary in building a house; it was a social critique that verged on conformist at the time. It was radicalism dressed in routine. Sometimes it seemed as if everyone was building a house, a perfect and socially sensible house, even if it in fact merely underlined the trends that existed, such as the trend toward building a second house. Thoreau's mother would have seen the house literature. She had helped draw up the plans for the new Thoreau house that Thoreau himself had just built; she had worked with the local carpenter. Thoreau's mother had probably been talking about villas, for example. Meanwhile, Emerson bought his land at Walden with the idea of building "a cabin or a turret there." Alcott built a house in a yard in Concord. Thoreau would soon establish a friendship with Daniel Ricketson, a literary wannabe, who built what he called "a shanty" alongside his house in New Bedford, Massachusetts. (Ricketson wrote Thoreau after reading *Walden* to tell Thoreau about his little house, and Thoreau got in touch with him two years later when he was lecturing in New Bedford and needed a place to stay.) As the country was being reshaped by the beginning of the suburbs, as farmers were replaced with commuters or with no one at all, country houses were a craze, and there was even a book about living in a simple house, built off in the woods, by Joel Tyler Headley, who, from 1847 to 1848, went to the

Adirondacks and described his time in a book entitled *Life in the Woods.* While writing at *his* retreat, Headley met failed investors who had been ruined by the Panic of 1837, an early and unintentional back-to-the-land movement, a group of former businessmen who were forced to live a life in the woods *à la* Thoreau.

Think of Thoreau as a trickster, a joker with subtle word-draped jokes, who, yes, would die at a comedy club today but charm and amuse a book group, who would be billed on a talk show as a man with an *obsession* for simplicity. (*Obsession* is a word that has replaced *passion* over the years, especially in publishing.) In this light, the house craze makes Walden, on one level, a humorously exaggerated imitation of the pattern books popular in his time. If we were living in the years before the Civil War, the humor would certainly be more obvious, and it might even be hitting us over the head. Here is the title page of John Claudius Loudon's *The Suburban Gardener and Villa Companion*:

COMPRISING

THE CHOICE OF A SUBURBAN OR VILLA RESIDENCE,

OR OF A SITUATION ON WHICH TO FORM ONE;

THE ARRANGEMENT AND FURNISHING OF THE HOUSE;

AND THE LAYING OUT, PLANTING, AND GENERAL MANAGEMENT

OF

THE GARDEN AND THE GROUNDS;

THE WHOLE ADAPTED FOR GROUNDS FROM

ONE PERCH TO FIFTY ACRES
AND UPWARDS IN EXTENT;

AND INTENDED FOR

THE INSTRUCTION OF THOSE WHO
KNOW LITTLE OF GARDENS AND

RURAL AFFAIRS

AND MORE PARTICULARLY FOR THE USE OF
LADIES

The ladies were responsible for the setting up of the residence, now that housework had become a specialized labor. In contrast, Thoreau was here taking on all tasks, building a house fit for a perch on a space of land that was fit for a villa.

In his "Estimate in Detail," John Claudius Loudon lists 181 items, because cost lists were de rigueur in house pattern books at the time. Recall Thoreau's famous list of costs for his house at Walden, which total "$28 12 ½." The least expensive house in *The Architecture of Country Houses* costs roughly $330. In his parody, Thoreau is able to address the matter at hand, building a house, while alluding to the larger subject, getting a living, or life, especially at a time when people are striving for houses that by the Transcendental accounting might be considered more than necessary—a time when the machinations of the economy are kicking some people out of their houses and into cities or unemployment while offering other people country houses, a trade-off.*

At his lectures, he spoke with the authority of a how-to guide, of a contractor on a home-building cable TV show, but he had another authority as well, one that was based on an opposite: he was building by unbuilding. He was taking apart ideas. He was dissecting ideas like ownership and the whole idea of getting a living, or a life. He was dissecting, for instance, the idea of cost, which for him was more than the mere cost of the materials and had to do with the cost of time, the trade-offs involved between striving to make enough money to own a house, on the one hand, and having an actual life to shelter, on the other. "I give the details because very few are able to tell exactly what their houses cost," Thoreau would soon write in *Walden*. He added: "I speak understandingly on this subject, for I have made myself acquainted with it both theoretically and practically."

* Thoreau's estimate of total food expense in *Walden,* $8.74, is probably no coincidence. It is precisely the same amount as estimated in *The Young Housekeeper,* a book about gastronomic reform written at the time by Bronson Alcott's cousin: a vegetable and cereal diet that boasted psychological and moral reforms for economic purposes, and cost $8.74 for eight months.

Chapter 8

A PLACE TO WORK

THE BIOGRAPHY OF THOREAU probably shouldn't matter as much as it does. Why do we care where he ate dinner? Why do we care whom he ate dinner with or if he ate vegetarian? What does it matter if he walked to Boston or took the train? But here is the thing: it matters because in *Walden,* Thoreau is the "I," or appears to be anyway, and he is the "I" who seems to suggest a course of living, a practice. The "I" then proceeds to describe the time he spent practicing. And it's a difficult practice. It's not a practice that says to keep doing things the way you are doing them. On the contrary, it's describing why you might change your life. It is a sugar-coated pill in that it comes in a beautifully written and oftentimes funny package. But then again, it is talking about *your* life and suggesting that you might be better off in changing it. It's not so difficult to build a house—it can be done—but it's difficult to build a life, even more difficult to start fresh and build anew. If Thoreau had written a book about going to the pond and not changing in any way, a book about how pretty the pines look at sunset, we might not feel the need to investigate his whereabouts while there, or indict him—we might not want to paint him a liar, in other words. If you ask me, if we didn't think he might matter we might forget him altogether. As it is, I see him in newspapers all the time, referred to as the hermit, as the strong-willed individual, a little much to handle. I like to joke that it is good for him that he did not have children, a biographical detail some often chide him for. If he had children, and had paid attention to them as much as he did Emerson's kids, paid or unpaid, he might not have had time to write the book—but if he'd had much to say about how we raise our children, then he'd really be a pariah.

The answers to the questions about what he was actually doing at the pond begin with what he was trying to do, which was work, and with a description of the place, which was not what it sounds like. It might have more appropriately been called Walden Lot, or Walden Woodlot, or the Place by the Pond Where We Get Our Wood or What's Left of It. To reiterate, Walden Pond—or the land all around it, since Thoreau's little house was not actually *on* the pond—was a woodlot, the place where the townspeople turned for wood, which was, aside from being necessary for shelter, a source of energy. Wood was their fuel, their coal, oil, and gasoline all in one. Wood was so valuable ministers were paid in cords of wood. Walden Pond was like an electric power plant or a gas station. Wood was also more and more scarce, since a lot of it had been used in the construction of the railroad along Walden Pond. Walden today is covered by woods, but these woods grew up as the human depopulation that started in Thoreau's time continued. While Thoreau was there, the woods were mostly being cut down. The woods were being cut down by the villagers, by the Irish railroad workers building shacks for themselves, and by the railroad for the railroad's sake. The woods were being cut down by Thoreau, who used pine to make his house, which was 10 feet by 15 feet, or 150 square feet.

He borrowed an ax in March (from Alcott, though years later, after the act took on significance, various people claimed it was from Emerson or Hawthorne). In April, he bought a shanty from an Irish worker, James Collins, to use the boards. "James Collins' shanty was considered an uncommonly fine one," Thoreau noted in his journal. (While Thoreau wasn't looking, another neighbor stole the nails.) He laid the planks out on the ground to bleach in the sun. He used a wheelbarrow to carry supplies to his site. He hosted a house-raising party to lift his little walls in May, using the old country custom, the kind of thing that polite and progressive Concordians frowned on—rude were the old ways, rude and unproductive. At the party, he gathered a group of philosophers and farmers and even had recent Irish immigrants on hand, all his Concord neighbors. He was doing what the Associationists had not been able to do: mix so-called high and low society, bring the mind of Philosopher Emerson to work side by side with the hands of Farmer Hosmer, if just for a few hours. "No man was ever more honored in the character of his raisers than I," Thoreau would later write.

He moved into his house on the Fourth of July. The experiment was a project of patriotic independence, harkening back to Jefferson's declaration. He announced his arrival at the pond in his journals. "I am convinced," he wrote to Horace Greeley a little while later, "both by faith and experience, that to maintain one's self on this earth is not a hardship but a pastime, if we live simply and wisely . . . It is not necessary that a man earn a living by the sweat of his brow, unless he sweats easier than I do."

Thoreau was poised to be the extra vagrant (his pun) at Walden Pond, the local wanderer taking his retirement near where they would one day put the Concord dump. House books praised the older home, the aged home, the true country house being the oldest house in the neighborhood, and Thoreau proclaimed his new house was just that, having a joke with the idea that there were no other houses in the neighborhood, even though there were a few Irish railroad workers still living there. At their peak, a year or two before, there was a shanty village, and women in Concord would send their children to tutor the Irish children, as alms for the poor. Emerson used to complain that the snow was colored with tobacco juice on his pond-side walks.

Moving to a house on a woodlot did not help Thoreau's reputation in town. He was already thought of as a ne'er-do-well, the woods burner who went to college and yet did not choose a career. Even pencil making wasn't good enough for him, people whispered. He seemed to like his reputation, as a matter of fact, playing the eccentric ne'er-do-well, teasing with pseudo-Socratic dialogue the villagers who teased him. "What are you doing?" someone might ask. "What are *you* doing?" he might shoot back. It did not help that he lived in the woods with people who polite Concord society might not have deemed as hard workers, much less upstanding citizens. Also living in shacks nearby were a drunk—the town barber, whom Thoreau often defended—and a freed slave. Alek Therien, the local woodcutter, was usually nearby. Aside from the kind of celestial soundtrack we might imagine when we read *Walden,* aside from the sounds of birds and of silence that Thoreau describes, the sound of axes chopping was ubiquitous in Concord and New England, and especially at Walden.

The first fall, he plastered the walls. There was not much to the interior, of course, and when you see re-creations of the house in Concord, it's a little smaller than a smallish storage place that you might rent in a city or in one of those vast storage facilities that you see when you drive the country via interstates. As far as furnishings went, he had a bed, a desk, three chairs, a kettle, skillet, and frying pan. The desk was a tall, table-like teacher's desk, the pine green paint around the keyhole scratched away—Thoreau kept his desk locked most of the time. It was the desk on which he presumably wrote the following description of the house, from *Walden:*

> My dwelling was small, and I could hardly entertain an echo in it; but it seemed larger for being a single apartment and remote from neighbors. All the attractions of a house were concentrated in one room; it was kitchen, chamber, parlor, and keeping-room; and whatever satisfaction parent or child, master or servant, derive from living in a house, I enjoyed it all. Cato says, the master of a family (patremfamilias) must have in his rustic villa "cellam oleariam, vinariam, dolia multa, uti lubeat caritatem expectare, et rei, et virtuti, et gloriae erit," that is, "an oil and wine cellar, many casks, so that it may be pleasant to expect hard times; it will be for his advantage, and virtue, and glory." I had in my cellar a firkin of potatoes, about two quarts of

> peas with the weevil in them, and on my shelf a little rice, a jug of molasses, and of rye and Indian meal a peck each.

He cooked outside the first year, and, after building a fireplace the second, cooked inside too. Aside from writing (at a pace that still makes a modern free-lance writer's head spin), he went to town to visit friends and family, to read the papers, to borrow books, to talk. Likewise, people were always visiting the cabin. The men working on the railroad stopped by. Ellery Channing stopped by, and even stayed there, sleeping on the floor, alongside Thoreau's cot, leaving his wife at home back in town. Occasionally, groups of people came to picnic at the cabin. He had a couple of dinner parties, writing the menu out in various languages. Emerson walked to Thoreau's house just about every day; he told Carlyle the woodlot that he had bought at Walden was "the best plaything I ever had." It was a twenty-minute walk from town. If there were a party at his parents' house or people over for dinner, Thoreau would stay late and be the last to leave. Walking back to the pond at night was no problem. He would walk along the railroad track until he came to the trail that led to his cabin, a trail made bold by his own wheelbarrow, his regular steps.

The first year Thoreau farmed beans primarily, as well as other vegetables—Thoreau was well-known in town for his skills as a gardener. He farmed on the land he cleared for Emerson, part of the barter that allowed him to use the land: Emerson wanted the land cleared of brambles so that he might plant pine, trees that could be later used as fuel, presumably. The second year, partly because of a frost that killed most of his crop, Thoreau would plant fewer beans and make money as a fence builder, painter, gardener. He would also take care of his woodchuck problem, by trapping the creature a second time and marching it two miles away.* He was undertaking two projects simultaneously: the "experiment," which was his life at Walden, and the project of writing, to finish *A Week on the Concord and Merrimack Rivers,* which he had already sketched out and seemed to be dying to rip into. He was rigorous in his attempt to keep his lifestyle simple, and this rigorousness was complementary to his literary productivity: the primitive life was the procedure, and the result the regular writing life, a life that he had attempted in New York City and wasn't able to achieve living at Emerson's house in Concord, due to his responsibilities as handyman, tutor, and nanny.

The experiment, meanwhile, would allow him to practice his version of an alternative life and moral reform. It was his practical utopianism, a new self-culture based, in part, on the culture of old, the culture before the factories and the establishment of the new middle class. "It would be some advantage to live a primitive and frontier life, though in the midst of an outward civilization, if only to learn what are the gross necessaries of life and what methods have been taken to obtain them; or even to look over the old day-books

*He had trapped the woodchuck once, but it returned. His friends joked about how much time Thoreau spent wondering aloud about whether he should allow the rabbits and the woodchucks to eat his beans or whether he should fight them. Remember, it was an experiment; Thoreau was on camera, in a way. Finally, he decided to trap it. He caught the woodchuck, released it, and was startled to see it return. His friends, all fishing one day, decided together to, in the words of one, "knock his brains out." (They were talking about the woodchuck.) Thoreau admonished his friends, saying the woodchuck had rights, had been there first, that they were "Squatters Sovereigns." They all laughed, given that in other instances, Thoreau was a hunter. Thoreau trapped the woodchuck again, marched it off two miles, and, according to Joseph Hosmer, who got a big kick out of all this, "never saw him more."

of the merchants, to see what it was that men most commonly bought at the stores, what they stored, that is, what are the grossest groceries," he wrote in *Walden*.

One of Thoreau's favorite things to do, when in town, was to comb through the old books of merchants—as much as an amateur naturalist, he was an amateur economist and demographer. He could see how our needs had changed even as our physical constitutions had not. The habit reminds me of a set designer I know who, for a film she was working on recently, was re-creating homes from the 1960s and marveled at how few things the average household had at the time—in the area of home electronics, just to begin with—compared to the average home now. "For the improvements of ages," Thoreau went on, "have had but little influence on the essential laws of man's existence: as our skeletons, probably, are not to be distinguished from those of our ancestors." A life improved, in other words, still had to be based on getting a life.

AT THE POND, AS MENTIONED, Thoreau got to work. Though he was assumed to be shiftless and lazy for running off to the cabin—for rejecting the usual work habits of, for instance, the college-trained, much less the industrious, citizen—he worked hard, his two years at the pond being, from the standpoint of a writer, tremendously productive. Aside from building a house and farming, he wrote a close-to-finished draft of one book, *A Week on the Concord and Merrimack Rivers,* and a first draft of what would become his masterpiece, *Walden*. In the meantime, he entertained numerous visitors, and—as the narrator of *Walden* states—visited the town frequently and regularly. He bathed in the pond, as he enjoyed bathing in streams and water all around Concord: "I get up early and bathe in the pond—that is one of the best things I do—so far the day is well spent," he wrote in his journal. He often bathed nude, a predilection that might not have been such a big deal a generation before but in Victorian America helped exacerbate his reputation as a weirdo.

He took long walks into the woods and through the fields. He did odd jobs, for different people, though many for Emerson. He built a fence for Emerson and got paid $5; he built a drain and laid a cellar floor for him and then painted a room. He fished for dinner in the pond, or cooked food from the stores in his basement. He surveyed property for farmers. The first fall, he built a chimney for himself. When he plastered his walls, he lived with his family for a few days while the plaster dried. In addition to working on his books, he wrote an article on Thomas Carlyle; Carlyle, like Emerson, wrote about Representative Men, men (almost always) chosen from the past and profiled so as to highlight moral traits and aspects of character. Thoreau considered this to be a form of hero worship. Additionally, Thoreau thought it avoided what Thoreau called "the Man of the Age, come to be called the working-man." Thoreau began to avoid the word *heroic.* Instead, he used *wild,* thinking of the wild man as himself, the new anti-hero. Now, with tongue planted in cheek, he could describe himself in his Walden state as a descendant of the heroic race of men: "I too sit here on the shore of my Ithaca, a fellow wanderer and survivor of Ulysses." It was a joke, but it was one of those jokes that he would have taken very seriously.

Also while at Walden, he spent a night in jail, an event that he would, after leaving the cabin, use in creating what is his most widely read essay, "Resistance to Civil Government," later retitled "Civil Disobedience." He did not write the essay, however, until *after* he left the pond.* On August 7 of the first year, the *Concord Freeman,* a local paper, covered the annual meeting of the women in Concord who were against slavery, a meeting convened on the anniversary of the freeing of the slaves in the West Indies. The meeting was held at Thoreau's house at Walden. For two weeks in the summer of 1846, he traveled from Walden Pond to Maine, to climb Mount Katahdin, and subsequently, at his desk in the cabin, wrote a long travel piece on his excursion—the beginning of several installments on Maine that he would, on his deathbed, put together as a book.

From his doorway, and all around the pond, he enjoyed watching the ice freeze and melt, and then during his stay, an ice company from nearby Cambridge, Massachusetts, bought ice rights from Emerson and the railroad, the landowners at Walden Pond, and began to harvest ice. Ice harvesting made the pond akin to an outdoor factory, like a commercial farm operation. Dozens more Irish immigrants arrived, as day laborers. They carried pikes to break out the ice, and stacked the blocks in straw along the pond for storage until they were shipped off by Irish laborers on the railroad. Emerson saw the ice business's arrival as a sign that his investment in property at the pond would pay off. "I am not without a prospect that my woodlot by Walden Pond will get an increased value soon; as Mr Tudor has invaded us with a gang of Irishmen & taken 10,000 tons of ice from the pond last week," Emerson wrote to his brother in Staten Island. Thoreau offered his cabin to the Irish workers who fell in the water and needed to dry out.

*The essay was not widely read at the time, and a review in a Boston paper suggested he should "take a trip to France, and preach his doctrine . . . to the rest of the red Republicans."

Thoreau then began his own ice harvest, a long project of taking the temperatures of ponds and rivers in the area. At the pond he was discovering this new kind of work: he could survey for himself, detail his environs for no reason other than his interest, a bottomless task.

Thoreau also used the ice as an opportunity to survey the bottom of the pond, drilling holes through the surface and dropping weighted lines—to an accuracy that holds under modern surveyors' review. He plays on the idea of the pond as a symbol of the soul in *Walden*.

> The greatest depth was exactly one hundred and two feet; to which may be added the five feet which it has risen since, making one hundred and seven. This is a remarkable depth for so small an area; yet not an inch of it can be spared by the imagination. What if all ponds were shallow? Would it not react on the minds of men? I am thankful that this pond was made deep and pure for a symbol. While men believe in the infinite some ponds will be thought to be bottomless.

He also collected fish and turtle samples for Louis Agassiz, the Swiss-born naturalist and professor of zoology and geology at Harvard. He met with Emerson and Alcott and their associates to see about starting another journal, a successor to the *Dial.* Mostly, though, he worked on his long river book, *A Week on the Concord and Merrimack Rivers.*

HE ALREADY HAD TWO HUNDRED PAGES OF NOTES toward *A Week* when he had arrived at the cabin, and a good idea of how it might go: a river story, a travel narrative through nearby nature, in the style of elegy, an elegy dedicated to his late brother. It would be a boiled-down seven-day version of his trip with John in 1839 into the White Mountains, a popular destination for landscape painters at the time, and a soon-to-be popular site in the burgeoning tourism industry. All told, the book took Thoreau over ten years to complete, but the bulk of it was done at Walden Pond, the obsessive rewrites continuing after he left, even after publishers became interested. The river trip would be augmented—or padded, in the minds of critics who did not like it—with Transcendental musings. It is a kitchen-sink book in that he managed to throw in pages and reworked pages from his journal as well as poems already published in or rejected by the *Dial* and snippets of previous writings.

Thematically, *A Week* goes from a "dead" river, the slow and sluggish Concord, to a "living" one, the fast-flowing Merrimack, which finally leads him to the White Mountains, the river's source. Rivers are the theme, and the idea of consciousness as a river-like entity is the philosophical current that runs through the work. "And our thoughts flow and circulate—and seasons lapse into the current year," he writes. Thoreau played tricks with time—a battle with Indians seems to take place in the past and then it is deliciously contemporary. The tricks oftentimes show how skillful a writer he was becoming, and oftentimes are confusing. Each chapter is a day of the week and has a theme, and very little of the actual boat trip is mentioned. The first chapter is a meditation on labor, for example, the second on Sabbath restrictions, which leads to a discussion of the books that might take the reader on trips to higher spiritual plains. He works in his trip to Wachusett, the mountain he calls the "worthy brother," with a long essay on friendship. Friendship is immortal, beyond death, the final resting place of John, the brother who is at the center of the book and yet never named.

There are brilliant strokes: that the book ends on silence, for instance, as if a sun had set, a week—or a life—over: "Silence is the universal refuge, the sequel to all dull discourses and all foolish acts, a balm to our every chagrin, as welcome after satiety as after disappointment." There are Thoreauvian literary raves: "But Chaucer is fresh and modern still, and no dust settles on his true passages. It lightens along the line, and we are reminded that flowers have bloomed, and birds sung, and hearts beaten in England." He writes on music and friendship, and on love in a way that sounds more like Thomas Merton, the Catholic writer, poet, and social activist, than Emerson: "Ignorance and bungling with love are better than wisdom and skill without. There may be courtesy, there may be even temper, and wit, and talent, and sparkling conversation, there may be good-will even,—and yet the humanest and divinest faculties pine for exercise. Our life without love is like coke and ashes. Men may be pure as alabaster and Parian marble, elegant as a Tuscan villa, sublime as Niagara, and yet if there is no milk mingled with the wine at their entertainments, better is the hospitality of Goths and Vandals." Essentially, though, the book doesn't work. It's got too much contemplation and not enough action; he doesn't manage to synthesize his Transcendentalism with a journey, the way he had in his most successful magazine essays. In a way, he worked too hard on it. It is the kind of book that every young writer thinks he ought to write—or put together—when in fact, he probably shouldn't. He sent it to publishers just before he moved out of the cabin, in the fall of 1847.

Then again, Thoreau was always his own toughest taskmaster; he would easily dismiss a recommendation by Emerson—to write an account of his college years, to profile a famous English writer—but when it came to his own self-assigned projects, he would pour his entire writerly being into them, never let up. If he ever had doubts about *A Week,* which he often did, it was always about something a rewrite, a new layer, might fix. As he wrote and rewrote *A Week,* he began keeping notes on his life at the pond, and eventually, just as he might go off to Emerson's house or call on Alcott and read aloud to one of them from *A Week*-in-progress, he decided to read from his notes on his life at the pond as a public lecture.

THE PUBLIC, THOREAU DISCOVERED, ENJOYED hearing him talk about his "experiment" at Walden, and, it's safe to say, he enjoyed them enjoying it. Hawthorne used to talk about how much Thoreau loved praise. People laughed when he gave his lecture—what the Lyceum record keeper called the "History of Himself," the lecture that would, over the course of the next few years, become *Walden*. Why were they laughing, especially in light of our contemporary opinion of Thoreau, which is solemn?

The fact is, Hawthorne wasn't the only one who saw him as a humorist. Among other things, Thoreau was joking. It's difficult to see now, partly because of our time: we live in the gold-plated era of memoir, where histories of selves have whole sections in bookstores, where the quotidian is pitched as poetic and winds up prosaic. But it's also difficult because the half-life of humor is often less than that of seriousness; jokes that are going to die—and he has a number of them—are going to die fast and decay quickly.

I'm not arguing that Thoreau was like a straight-ahead comic you'd see today on cable TV ("Thank you, thank you—I'll be at the Concord Lyceum all this week!")—but he was a little like one, in fact, with a patter based on financial papers, for example. Think of his later lecture, "Life Without Principle," in context with investment mania then (or now), *principle,* as in *belief,* being a pun on the financial kind. Thoreau was, naturally, using humor to make some very serious points. But he was also—especially in *Walden,* and especially in the idea of Walden, and in the first chapters (and as opposed to the tone of *A Week*)—kidding around, and he thought kidding was important. "We think the gods reveal themselves only to sedate and musing gentlemen," Thoreau wrote. "But not so; the buffoon in the midst of his antics catches unobserved glimpses, which he treasures for the lonely hours."

Thoreau understood what reformers and serious thinkers of all kinds often forget: that humor helps. Transcendentalism, Thoreau argued, "needs the leaven of humor to render it light and digestible." He leavened his message about the economics of everyday life with puns and wordplays and outright one-liners. His was a wry humor, the always-composed tone of that iconic frontier comic, the famed Yankee peddler. The peddler was a storyteller, a loner, a wandering satirist, with a wise eye to the ways of the people he was passing by; the peddler was known to answer a question with a question, as *Walden*'s narrator announces that the book is itself a response to the neighbors' wonderings: "Some have asked what I got to eat; if I did not feel lonesome; if I was not afraid; and the like." So begins a long rhetorical interrogation. "Who made them serfs of the soil?" the narrator asks of the overworked farmers in town. "Why should they eat their sixty acres, when man is condemned to eat only his peck of dirt? Why should they begin digging their graves as soon as they are born?"

At the heart of the joke is Thoreau's "I," the "I" being not just Thoreau but Thoreau plus. The development of the "I" is no small feat, a key ingredient for any writer who dares to try his hand at the first person. Many people use "I" when writing, of course, but few develop it. If you refine the "I," if you stylize it, you end up with an "I" that is, as far as literary genetics goes, more related to personal essays than to, say, travelogue.) As a writer, you work on your "I"—tone it up or down on occasion or, conversely, push it—and when you succeed, you develop an "I" that is true to you, as opposed to being truly you, a crucial distinction. If the "I" were truly you and you were the author, the reader would endure something along the lines of mere transcription, a bore. The successful "I" is not false, but it must be what the "I" in Walden claims to be—*extra.* The extravagant you, the extraordinary author, who is, in life outside the book, likely to be ordinary, due only a peck of dirt, unless from a place unheard of, or from another planet. The author of *Walden* is different from the "I" in *Walden,* even if they are the same person.

With a wink and nudge and nod, Thoreau used the Yankee "I" that he had made his own, especially in his various plays on words: "In any weather, at any hour of the day or night, I have been anxious to improve the nick of time, and notch it on my stick too; to stand on the meeting of two eternities, the past and future, which is precisely the present moment; to toe that line," he would write in the opening section of *Walden*. *A Week* was a step backward, given what he'd learned in New York; it was a pure Transcendentalism, Thoreau doing Emerson, in book form, mostly serious. But the "experiment," and the *Walden* it would ultimately yield, was by the Thoreau who had written the genial essay "The Landlord," the free-lance who had to sneak his philosophies in behind a more popular front line. People wanted to get ahead, financially speaking, in the 1840s and '50s, and here was his take on the popular theme, apologizing, in jest, for his irrelevance: "Perhaps these pages are more particularly addressed to poor students. As for the rest of my readers, they will accept such portions as apply to them. I trust that none will stretch the seams in putting on the coat, for it may do good service to him whom it fits."

Just joking was also part of another particular cultural trend. At the time of the experiment at *Walden,* and its subsequent publication half a decade later, antebellum America was in the midst of a national obsession with grammar, with word roots and origins. It was part of the trend to reform, to self-improve, and the same surge in printed matter that was fueling the trend in home and garden literature was also fueling a love of games and spelling books, puzzles and word histories. In addition to parlor games of wordplay, spellers flooded frontier schools, where wisdom, it was argued, was attained through spelling: "the youth from lisping A.B.C,/ Attains, at length, a Master's high degree." Popular manuals, such as *The Scholar's Companion; or, A Guide to Orthography, Pronunciation, and Derivation of the English Language,* were huge best sellers, as were books that collected puns, jokes, and wordplay. "I feel an antipathy towards a whale because it has a tendency to blubber," went a line in the *American Comic Almanac.* In the year that Thoreau lived at the pond, the *Concord Freeman* ran an advertisement, by E. Gunnison, the village cleaner: "To Dye, or not to Dye," it said, adding, "he is yet alive, though often dyeing! And is ready to dye for anyone who may need his services in the DYE-ING ART." Etymology was in vogue, in all seriousness and silliness. In the history of words was the past, the thinking went. The Transcendentalists weren't the only ones who thought words brought us closer to original thought, to the divine. For instance, in addition to being a preacher and minister, Joseph Smith was a translator, translating the Book of Mormon from a revealed ancient text in 1830 as only he knew how, a divine etymology.

Thoreau was passionate about the roots of words—he owned seventeen dictionaries—and he believed that good writing not only used words in their historic meaning but brought the reader to the deeper wisdom that the words themselves contained. A smart pun was not just a pun, but a pun that directed the reader to a truth, sometimes tripped or tricked him into it, sleight of word. Language was itself spiritual, in other words, an Emersonian idea, words clues to a life truly lived. Whereas the average punster might pun for the pun's sake, Thoreau looked for the pun that would, in Thoreau scholar Sherman Paul's words, "drive to the radical meaning of things." In 1853—near the time he was writing the last drafts of *Walden*—after reading Richard Trench's *On the Study of Words,* with its often specious word roots, Thoreau notes Trench's explanation of the word *wild*—which, as Trench has it, is derived from *willed:* a wild man is a willed man.

> A man of will [is one] who does what he wills or wishes, a man of hope and of the future tense; for not only the obstinate is willed, but far more the constant and persevering. The obstinate man, properly speaking, is one who will not. The perseverance of the saints is positive willedness, not a

> mere passive willingness. The fates are wild, for they will; and the Almighty is wild above all, as fate is.

Thoreau wanted to be a wild man, but he wanted his wildness to come from inside, from his own free will, as opposed to the wild of the land.

THE WORD *EXAGGERATE* IS DERIVED FROM THE LATIN *exaggerare*, meaning *to heap up,* and more than anything, Thoreau was an exaggerator, literally and literarily, digging out a basement next to a pond and then heaping on associations, piling up meanings hidden and direct in the literary work that followed—lively jokes and allusions assembled like the heaped-up muskrat's den. Despite decades of solemn readings, despite our desire to read *Walden* literally, as a withdrawal to the faraway woods, Thoreau was in fact the comedian grabbing his audience by the collar, a con man with confidence, intending to inspire alternate meanings. "I desire to speak somewhere without bounds; like a man in a waking moment, to men in their waking moments; for I am convinced that I cannot exaggerate enough even to lay the foundation of a true expression," he writes in *Walden*'s conclusion. He plays with the word *Walden* itself, punning at one point on the Waldenses, a sect of religious dissenters founded by a twelfth-century French merchant, Peter Waldo, who rejected the authority of the pope. ("On the Late Massacre at Piedmont" is a sonnet by John Milton about their persecution in the 1600s, in the Alps.) In a less public moment, Thoreau wrote to his friend Harrison G. O. Blake: "I trust that you realize what an exaggerator I am,—that I lay myself out to exaggerate whenever I have an opportunity,—pile Pelion upon Ossa, to reach heaven so. Expect no trivial truth from me, unless I am on the witness stand. I will come as near to lying as you can drive a coach-and-four."

Shortly before he died, Thoreau would worry he was too wedded to using opposites in his style, that perhaps Emerson had been right about scolding him for it. "The habit of a realist to find things the reverse of their appearance inclined him to put every statement in a paradox," Emerson said. This didn't just upset Emerson; it has confused a lot of readers. But it was Thoreau's style, to turn something inside out, to stretch it to its extreme, in a way that strict interpretations might deem false or, worse, lies, that would contribute to our general misunderstanding of Thoreau. But for Thoreau, in addition to being fun, exaggeration was a way to truth. In his journal, in 1851, he talked about life as "an experience of infinite beauty, on which we unfailingly draw, which enables us to exaggerate truly." He used a comic tone that his audience wouldn't have missed—what Constance Rourke, in her 1931 classic study, "American Humor," called "an unmistakable native authority, the true sound of native speech."

"You will pardon some obscurities, for there are more secrets in my trade than in most men's, and yet not voluntarily kept, but inseparable from its very nature," the narrator declares in *Walden*. "I would gladly tell all that I know about it, and never paint 'No Admittance' on my gate."

One evening, during his year at the pond, he stayed late in town to give his latest lecture, "History of Himself." It went well. The Brook Farm community heard about it and showed up the next week when he did a repeat performance. "Mrs Ripley & other members of the opposition came down the other night to hear Henry's account of his housekeeping at Walden pond, which he read as a lecture," Emerson wrote in a letter to Margaret Fuller, "and were charmed with the witty wisdom which ran all through it all." Note that Emerson calls them the opposition. Note that the opposition loved Thoreau.

LIVING AT THE POND WAS THE FIRST JOKE, the one he used to build the book. As an act of parody, the "experiment" alluded, first of all, to the history of land and manhood, of even citizenship in America. In *Letters from an American Farmer,* published in 1782, J. Hector St. John de Crèvecoeur defined the new American man as a father and a landholder. Thoreau, meanwhile, built his house as a single man who has no land, no servants, hired or otherwise. And in choosing a crop, he chose beans. Especially nowadays—when food choices can seem to describe personal choices and crops can seem to have almost theological implications, some of them tied to issues involving the sustainable use of the land, some to personal health—eyebrows are raised at the notion of Thoreau planting beans. (Some modern plant specialists question if the choice was environmentally sound.) But let's take a look at these beans.

On one level, the choice to farm beans was a critique of farming methods. With self-sufficient farming declining, New England farmers were competing with the high-yield farms to the west, with farmers who sent their grain to Boston by canal and train. New England farmers pushed their fields to productive extremes, to the detriment of the soil and the farmers' own economic health. Thoreau began with what would be considered poor pond-side soil—"One farmer said that it was 'good for nothing but to raise cheeping squirrels on'"—and planted beans. Then, with a crop worth nothing, literally and metaphorically, Thoreau proceeded to gross $23. Twenty-three dollars was not nothing in 1847. He reports his profit as

8.71½, the ½ a sarcastic swat at the obsession with financial detail, and in profiting at all, he proves the point that he can make a profit without killing himself or the land. He could have a life, in other words, for beans—a little something for nothing. "All things considered, that is, considering the importance of a man's soul and of today, notwithstanding the short time occupied by my experiment, nay, partly even because of its transient character, I believe that that was doing better than any farmer in Concord did that year."

On another level, beans are a joke. Already, in antebellum America, in the slang of the time, beans are worth nothing—*beans!* Beans criticize themselves, announcing themselves in the horn honk of flatulence—*beans!*

Meanwhile, Thoreau's activity at the pond—his "housekeeping," as Emerson termed it—was another reference, this time to another great middle-class trend of the day: the housekeeping and home economics manuals that were wildly popular, chief among which was Catharine Beecher's *A Treatise on Domestic Economy.* Published nationally, it went through three editions and fifteen reprintings, beginning around the time that Thoreau had first moved into Emerson's house. It was used as a textbook by Massachusetts schools. It was a big best seller, in other words. In it Beecher addresses many of the topics that Thoreau ostensibly raises—health, eating, shopping, housework. Food—or what food to eat—was a growing issue in America as it is today. The narrator of *Walden* is not so much the radical vegetarian—as anti-environmentalist critics of Thoreau like to say—when viewed alongside a health reformer like Beecher, who suggested Americans refrain from meat. (Thoreau ate meat on occasion, if politeness compelled him to do so when eating with nonvegetarians, what people now sometimes call a flexitarian.) He was not the crazy teetotaler—the critic of Thoreau as pious monk often hits this note—when viewed alongside the public's acceptance of Beecher's admonition to reject "stimulating drinks." "As a nation, the Americans are proverbial for the gross and luxurious diet with which they load their tables," Beecher writes.

Thoreau and Beecher were talking about the same themes but for completely different reasons. Beecher saw America as an endpoint of democracy, and painted housework and domestic economy as a patriotic means of ensuring the successful American enterprise. Thoreau, meanwhile, called patriotism a maggot in the head. Beecher advised American women to see housework not as drudgery but as a domestic "business," and she argued that European women of means were less engaged in the life of their nation because they did not work. The backdrop to this advice concerned the labor pool. There were not enough women in the United States for them to be employed as domestic servants in 1843. Two things would soon change this situation: the arrival of hundreds of thousands of lower-class Irishwomen and the growth of cities, which fueled middle-class income while causing a drop in the lower-class income, the perfect storm for the beginnings of a large new domestic-labor market. The change would happen nearly in the space of a decade, but in the meantime, Beecher justifies American housework, once a shared familial responsibility, as a task for women that shall bring on "the regeneration of the earth." The men will be free to work long hours in factories, as clerks, for instance, in the perpetuation of "enterprise." "It is the building of a glorious temple, whose base shall be coextensive with the bounds of the earth, whose summit shall pierce the skies . . ."

The dishes, cleaned by women, would lead to a bright

future for America; domestic work would clear the nation's plate for the hard stuff. But Thoreau was looking at domestic labor in an older sense, *à la* John Donne in 1644: "The knowledge therof is so domestique, so neare, so inward to us, that our conscience cannot slumber in it, nor dissemble it." Thoreau was looking to unlock our inner housecleaner or, to use another word often associated with home workers, our inner domestic, the word *domestic* being rooted in *house,* in Latin *domus*. "Our lives are domestic in more senses than we think," Thoreau wrote in *Walden*.

As magazines about luxuriously simple lives today compel us to buy an increasingly greater number of things that might simplify our lives, Beecher's *Treatise* actually compelled consumption, so as to promote the national economy, an argument still frequently heard. "Suppose that two millions of the people in the United States were conscientious, and relinquished the use of every thing not absolutely necessary to life and health," Beecher asks. "It would instantly throw out of employment one half of the whole community. . . . The use of superfluities, therefore, to a certain extent, is as indispensable to promote industry, virtue, and religion as any direct giving of money or time." Thoreau in his stay at *Walden,* in merely proceeding with his experiment, suggested "another alternative than to obtain the superfluities." Thoreau sought to engage the public using the lingua franca of reform movements. He was a philosopher speaking in the language of the self-help authors. He was twisting the recipes for reform—tainting them with Transcendental and even Associationist themes, writing a guidebook to a just society based on something more than the trendiest of trends, based as well on the "ruts of tradition and conformity" that we all end up driving into.

Much has been said about Thoreau's apparent aversion to women, as a lifelong bachelor, but it's not difficult to imagine that the wives of the various Transcendentalists he knew might have enjoyed his wit and noticed that he was a handy man—he was Emerson's au pair, after all. At the pond, he was doing so-called women's work, and calling it the work of humanity. And he wasn't saying just that everyone should do housework. Most radically, while Beecher attempted to change the perception of work from "vulgar" to "noble," Thoreau aspired to combine labor and leisure, to make vigorous work more like spontaneous play. "Idle time" was the bane of the new industrial economy. For Thoreau work was itself a pastime. In the roots of the word *labor,* Thoreau saw an opposite. The Latin noun means *work, labor, toil.* The Latin verb can mean *to slip* or *to fall away* or *to glide by*. In the "House-Warming" chapter of *Walden,* the narrator skates across the pond, collecting firewood as he glides joyfully. At a time when work and leisure were first becoming separated in America, Thoreau wrote that "men labor under a mistake," and the mistake he was talking about was that people were confusing work for work when it could also be play. They were becoming obsessed with making a living and forgetting to have a life.

BUT WHY DID HE CHEAT? WHY DID HE GO INTO town? Why did he leave the pond? These are the questions with which the poor guy is repeatedly battered. But if you ask me, the answers to these are built into the construction of the house at *Walden,* woven into the walls like so much lathe and plaster, the house itself being another example of Thoreau following the crowd to offer the crowd an alternative route, a detour off nineteenth-century America's main roads.

The humor of *Walden* is partly in its satirical structure; in a lot of places, it is like a parody of *Architectural Digest* designed to reassess home design altogether. What is the point, Thoreau asked, of having a house that makes us look like every other man? What is the point of calling a woman's house suitable merely for its resemblance to something seen in a newspaper or magazine, be it in the nineteenth or twenty-first century? Is the home you build and live in for the sake of you, the person inside, or for the sake of everyone on the outside? These are the questions that people often think they are on top of but often are not, really.

Thoreau's house, in its design and construction, was, as noted, a nod to the genre of home-building books that filled wealthy parlors at the time. The words that are maybe most famously attributed to Thoreau—"Simplicity, simplicity, simplicity!" the narrator shouts in *Walden*—are practically paraphrased from the house pattern books that were all the rage in his day. "Large estates," wrote Andrew Downing, the famous landscape planner of the day, "large houses, large establishments, [*sic*] only to make slaves of their possessors . . . It's so hard to be content with simplicity!" Once again, Thoreau was twisting a popular argument for reform; he saw the irony in the craze for "simple" houses (in many cases second simple houses) off on land that was until recently productive community farmland. In building his house, Thoreau was following the precepts laid out in the house-pattern literature, first of all, by staying close to town. The idea was to *retire* to the villa—which is why the act of moving to a small cottage was called *retirement*—but to keep in touch, though just a little removed. In pamphlets such as "Reflections on the Necessity and Advantage of Temporary Retirement," the writer called for the term of retirement to be short, along the lines of Thoreau's two-year stint, or even shorter. Wordsworth, who promoted the idea of villa retirement, described "a habitation in this peaceful vale" for what he called "an allotted interval of ease" in his *Poems on the Naming of Places.* In fact, there was what scholars call "a retirement phenomenon," and when one retired, one was not cheating when one visited one's friends. It was a genteel phenomenon, and an important aspect of it was never reliquishing social connections, never really going away.

The country house was to be a primitive thing, in Thoreau's mind and in the minds of pre–Civil War America's trendsetters as well. The trendsetters were inspired by the architecture described by Vitruvius, the Roman architectural writer and military engineer—again, the Romans loomed large! It was a classical primitiveness that progressives of the day believed would be the setting for a new American golden age. The progressive home was to be built on a hill; Thoreau constructed his on the burrow of a woodchuck (a joke that even his buddy Ellery Channing failed to get). Ruins, even faux ruins, were a necessary accoutrement to the well-built American retreat house; thus, Thoreau constructed part of his house with the ruins of a shanty. The pattern books called for the houses to be surrounded by native plants: "no dress trees," one admonishes. The idea of using native plants about his house seems to square easily with Thoreau as popularly understood, an ecological radical. But it was not radical. He found the same advice in the suggestions of his contemporary architects and gardeners. What Thoreau does, in part, is take the idea further, at least rhetorically, to test its strength. He cherishes the weeds, the not-so-romantic nature, the unkempt beauty. Often he writes of using sumac, for instance, a native plant considered obnoxious at the time, even by native-plant enthusiasts, calling it "pleasant though strange to look on."

Sometimes his commentary on his peers and their ideas was not veiled at all. Thoreau saw the value in the ideas behind the progressive architecture of his day, but saw the weakness of their underpinnings, that what they were really after—moral and societal reform—required a more solid foundation than just a nod to the picturesque. Downing wrote:

> If you would fix upon the best colour for your house, turn up a stone, or pluck up a handful of grass by the roots, and see what is the colour of the soil where the house is to stand, and let that be your choice.

And Thoreau, in *Walden,* wrote:

> One man says, in his despair or indifference to life, take up a handful of the earth at your feet, and paint your house that color. Is he thinking of his last and narrow house? [By which Thoreau means the grave.] Toss up a copper for it as well. What an abundance of leisure he must have! Why do you take up a handful of dirt? Better paint your house your own complexion; let it turn pale or blush for you. An enterprise to improve the style of cottage architecture! When you have got my ornaments ready, I will wear them.

Rather than be the same color and style as every other reformed house, the house might reflect you, its individual inhabitant. It's a question of moral architecture. If the house and the man were related, as the critics were proposing, then why not flip the prescription and, instead of remaking the house, remake the man? Like *The Picture of Dorian Gray* in reverse, the house might end up looking good too.

LIFE PROCEEDED SMOOTHLY FOR THOREAU during his two-year retirement at the pond; he enjoyed his parody of retirement, and when he returned home, he would sometimes ask himself why he ever left Walden Pond. Work and play combined; art and commerce were in friendly relations, for a while at least. "Both place and time were changed, and I dwelt nearer to those parts of the universe and to those eras in history which had most attracted me," the narrator of *Walden* reports. "Where I lived was as far off as many a region viewed nightly by astronomers." He was not far, of course, and that was the part of the joke that he was serious about: to be at a remove didn't mean you had to remove yourself. His neighbors probably engaged him more then and over the course of his life as a result of the experiment at Walden.

At the pond, the first 117-page draft of *Walden* took shape, the first chapter especially, as a riff on the idea of *economy*—economy being a matter on the lips of Concordians and Americans in 1847 as today and, I imagine, as tomorrow. The frost that killed his beans came and went, at which point Thoreau retired from farming for the most part. He began to make more notes in his journal about the pond and the woods in general. In the fall of 1846, he traveled to Maine, and climbed Mount Katahdin. Over the winter, he wrote a second draft of *A Week*. On April 8, he noted that the pond was completely ice-free. Things seemed to be winding down as far as feverish writing went, as he had begun to inquire with publishers about *A Week*. Around August, Emerson wrote in his journal that he had talked to

Thoreau, who had said he was "nowhere, doing nothing." That fall, after he had helped Alcott build a summerhouse for Emerson, Lidian Emerson asked Thoreau to return to the Emersons' while her husband was away in Europe. In a flash the Walden experiment was over. Thoreau sold his house at the pond to Emerson, who sold it to a farmer, Hugh Whelan, who took it away to a spot very near Walden Street today (the cellar Whelan dug still remains). The house at Walden would become, first, a shed, then part of a larger building. It was, as a matter of fact, recycled naturally, consumed into the fabric and subsequently the history of the town and the fields, without a plaque or a note, effortlessly. With *A Week* ready to send to publishers, Thoreau turned back to rewriting *Walden,* which he was going to do over and over again for the next six years.

Understandably, the publishers were hesitant about *A Week*. The Transcendentalism that Thoreau managed to sneak into his nature essays was in vast supply here, a questionable marketing tool. After leaving the pond, he finally found a publisher who would publish the book only if Thoreau would cover the costs of the book himself if it did not sell. Thoreau agreed. It did not sell. It was reviewed unenthusiastically at best, especially for its use of the first person, which Thoreau had not yet honed, though he was already having better luck with it in his lectures and with drafts of *Walden*. James Russell Lowell, in the *Massachusetts Quarterly Review,* wrote: "WE [*sic*] think that Mr. Thoreau, like most solitary men, exaggerates the importance of his own thoughts. The 'I' occasionally stretches up tall as Pompey's pillar over a somewhat flat and sandy expanse." (Pompey's Pillar was a landmark in what is now Montana for travelers on the Oregon Trail, western travel narratives being not only a practical read for people considering western travel as a career option but a popular armchair read as well.) Many people found *A Week* scandalous—"with reflections that may shock every pious Christian," as one reviewer wrote. The *New Hampshire Patriot* was more positive, and might have caused Thoreau, who read everything, including the popular press, to envision the *Walden*-in-progress as pleasing them more: "This is a remarkable volume and [its] author is a remarkable man. The title is very unpretending and gives but a faint idea of the contents of the work. Few men think as much as they should." One thing that ought to have set off alarm bells for Thoreau was the fact that Alcott liked *A Week,* calling it "fragrant with the life of New England woods and streams." Alcott's writing was often incomprehensible.

After *A Week,* Thoreau owed $300 to the publishers, easily several years' income for him. To get out of debt, Thoreau manufactured a large order of pencils. However, in the time that he had been at the pond, the pencil market had been flooded with less expensive high-quality German pencils. Thoreau was forced to sell his pencils at a loss. He tried next to sell cranberries in New York; again, he misplayed the market. Finally he had handbills printed, advertising himself locally as a surveyor, a profession he would first use to clear his immediate debt and subsequently rely on for the rest of his life for income and for allowing him to see all of Concord and its environs at once more closely and more profitably. He took the copies of *A Week* himself, moving them into his parents' house, 706 copies out of 1,000. "They are something more substantial than fame, as my back knows," he wrote, "which has borne them up two flights of stairs to a place similar to that to which they trace their origin." Thoreau could be haughty, of course, but he certainly knew how to be self-deprecating. "I now have a library of nearly nine hundred volumes, over seven hundred of which I wrote myself," he wrote to a friend.

And yet the failures of *A Week* cleared the way for *Walden,* the way he'd cleared the brambles from his land at the pond in order to grow beans. In some way, *A Week,* the elegy, was about finding a way to nature, or a life based on some higher truth, but in *A Week* Thoreau never actually accesses nature—it's a travelogue that never goes anywhere. "We were always passing some low inviting shore or some overhanging bank, on which, however, we never landed." He had forgotten a lot of the lessons he learned while in New York about writing commercially. With *Walden,* he would attempt to access nature more directly and, in so doing, find a way to live better. Sometimes writers have to write things to exorcise them. With *A Week,* Thoreau had gotten his pure Transcendentalism out of his system. As it happened, *A Week* ends on a Friday, a smooth sail, the week ready to begin again. (Thoreau especially enjoyed this section, and referred to it on his deathbed.) Now Thoreau was set to begin again, or yet again. One of the things he discovered while writing *A Week* was that he wanted to delve deeper into the idea of nature as salvation, as opposed to Christ. "I feel that I draw nearest to understanding the great secret of my life in my closest intercourse with nature," he wrote. He had gotten all mystical, living at the pond, at the same time that he had become more obsessed with transcribing the actual: the details of tree rings, the levels of rivers. The two year-stint at the pond had brought him to the definition of what he might have called a good life: "A life of equal simplicity and sincerity with nature, and in harmony with her grandeur and beauty."

THE REVIEWER OF *A WEEK* who perhaps most upset him was Emerson. Emerson had praised the book to Thoreau in person. Eventually, Thoreau realized that Emerson was not being honest with him. "I had a friend, I wrote a book, I asked my friend's criticism, I never got but praise for what was good in it—my friend became estranged from me and then I got blame for all that was bad,—& so I got at last the criticism which I wanted," Thoreau wrote in his journal. In *A Week,* Thoreau had followed the precepts of "The American Scholar." Now, *Walden* was going to be a different kind of book. A different kind of book required an end to the apprenticeship. Emerson, after all, liked the idea of a cabin, but thought Thoreau was taking the whole thing too far. Emerson was always telling Thoreau that life should be lived in that place between idealization and the limits of reality—what he called "the mid-world." "Henry Thoreau is like the woodgod who solicits the wandering poet & draws him into antres vast & desarts idle, & bereaves him of his memory, & leaves him naked, plaiting vines & with twigs in his hand," Emerson wrote in 1848 in his journal. "Very seductive are the first steps from the town to the woods, but the End is want & madness." Thoreau went from being described by Emerson as "my valiant Henry Thoreau" or "my brave Henry" to a writer who confuses meanings, who purposefully luxuriates in opposites, things that upset Emerson but become the hallmark of Thoreau's style.

Nevertheless, after living at the pond, Thoreau reported for duty back at Emerson's house while Emerson embarked on his European tour. Thoreau's time at the Emersons' appears to have strained the relationship more. Emerson claimed to value his domestic life, and surely he did, and yet he went away to Europe on an extended trip when his wife's health was frail, when his children were young. Emerson wrote depressing letters to his wife, even while she asked for more loving ones. "Well, I will come again shortly and behave the best I can," he wrote. "Only I foresee plainly that the trick of solitariness never can leave me."

Thoreau wrote privately of an unnamed woman whom Thoreau scholars presume to be Lidian. "I still think of you as my sister," Thoreau wrote. "I presume to know you. Others are of my kindred by blood or of my acquaintance but you are mine. You are of me & I of you I can not tell where I leave off and you begin.—there is such a harmony when your sphere meets mine[.] To you I can afford to be forever what I am, for your presence will not permit me to be what I should not be."

Obviously, Thoreau loved Lidian, and when he wrote letters to Emerson, he talked about Emerson's family in a way that, even if it was well-meaning, had to sting. "Lidian and I make very good housekeepers," Thoreau wrote. "She is a very dear sister to me. Ellen and Edith and Eddy and Aunty Brown keep up the tragedy and comedy and tragic-comedy of life as usual. . . . [Eddy, Emerson's son,] very seriously asked me, the other day, 'Mr. Thoreau, will you be my father?' I am occasionally Mr. Rough-and-Tumble with him that I may not miss him, and lest he should miss you too much. So you must come back soon, or you will be superseded."

When Emerson did return, he upset Thoreau by having picked up Brahmin tendencies—he smoked cigars after dinner, and he founded a new club to replace the Transcendental club. It was called the Town and Country Club, and even though Thoreau wanted to socialize, he was still very, well, *Thoreau*. The Town and Country Club seemed to him to be more about socializing and dinners. Thoreau found it all very British and refused to take part.

They were still friends, even if the mentor bond had been broken.

Their relationship changed, as it had to. It changed on Thoreau's side; he was even more bold with his mentor. It changed on Emerson's side: rather than barter a family-style relationship, he would return to paying Thoreau for tasks. Emerson finally realized that wages offered a more moral basis for relations of in-house labors.* Emerson still trusted Thoreau like no one else in his life. In 1850, when Margaret Fuller was returning from Italy with her husband and child, her ship hit the rocks off Fire Island, in New York. Emerson dispatched Thoreau to find her body and a manuscript. By the time he got there, the wreckage had been picked clean by scavengers.

* It was a realization that—as it did for a lot of northerners who were facing similar home-labor issues at the time—helped Emerson finally support abolitionism wholeheartedly, in fact. In "Emerson's Domestic and Social Experiments: Service, Slavery, and the Unhired Man," the historian Barbara Ryan writes, "Emerson had learned that wages kept masters masters."

Chapter 9

IMAGINE A CITY

I SUPPOSE THIS WOULD BE a good time to explain, not that anyone asked, that I fell backward into *Walden,* as a writer writing a book about another place entirely, the swamps of New Jersey, just hoping to make a point, and hoping to make a point with the help of the popular Thoreau. I had read *Walden* in high school, or parts of it, and I had the idea that most everybody has: guy in a cabin living alone in a faraway place, a freak, possibly, albeit pious. I tried to make a little joke using his checklist for living expenses, and it was more difficult than I had anticipated. The problem, I eventually realized, was I was trying to make a joke with a joke, which is like trying to make a sandwich with a sandwich: the second time, quality suffers. The best parodies end up somewhat parody-proof, and here I was trying to make humor with something that I thought was serious and factual when in fact Thoreau, in the opening pages of *Walden,* verges on spiritual slapstick.

A couple of years later, still not a whole lot wiser on the subject of Thoreau, I wrote another book, this time using *Walden* as my structural centerpiece, taking it more head-on, mimicking its structure, not that anyone noticed. I was counting on people's ideas of the Thoreau you know: the eccentric who went into the wild to live monastically. My topic was unusual, I'll admit, and not terribly natural-seeming.* But I was trying to have some fun with the concept of nature, looking for where in the sand was the line between what was natural and what was not. In so doing, I was thinking of *Walden* as the ultimate Nature Book, which meant that before I wrote my book I went through Thoreau's looking for places to hang up some commentary, which in turn meant I spent some time banging on the walls of his book, searching for the solid beams, the studs.

In my own experiment, I set up a little platform for nature viewing in the middle of the city, New York, where I was not a mile from my neighbors but in the midst of them, millions and millions of them. Naturally, I found that people thought I was a little extravagant, a touch radical, by virtue of merely looking for nature in a city, and thus I felt a little separate myself, even though I am no different from anyone else, in innumerable ways. Who is separate? Who is strange? And what is radical? I have to admit the older I get—and as I was typing this book I just passed the age that Thoreau was when

* Okay—I didn't want to say it, but my topic was about rats. I spent a year in an alley—an alley being, by definition, walled-in, pun intended—keeping an account of the life of some rats.

tuberculosis finally killed him—the less radical *radical* sounds to me. Indeed, the etymologist in Thoreau has always pleased me, and I realize that word games are in my American genes, leading me to note here that the roots of *radical* are just that: in Latin, *radix* means *root*. And to get radical means, in one sense, to get back to the first principles, the original meaning, the roots of a thing.

This is *Walden,* as far as I can tell. It's an implement that clears away the topsoil—some good, some tired—and offers a view of the roots, a view that was maybe not as radical as we tend to think, and nowhere near as conservative as the popular conception of Thoreau might contend. It's a work of virtuosic simplicity, a work that tears down to reconstruct. But again, it's a place. The great problem with *Walden,* as it reads to me today, is that we think of it as a prescription—how many newspaper columns have I read over just the past few years that have mentioned Thoreau in relation to giving up a car or giving up a cell phone or adopting a particular diet?—when, in fact, it's a work of art, a public art piece. It's a place that Thoreau crafted with the intention, I would argue, of inspiring you to build your own.

After living at Walden, the pond, Thoreau had to write *Walden,* the book. Back in town, Walden stopped being a place that he had been to and became the place he created. He intended *Walden,* the book, to have broad appeal, to please his readers, to amuse them, as well as do other things to them—Thoreau had trained to be a preacher and, like Emerson, he was one in the end. He was working in the culture, not apart from it, and the culture was the culture of enterprise, as in business. In this place called Walden, Thoreau transformed his stay at the pond into a business proposal, the business an enterprise of himself. The words of enterprise began to flip-flop, as Emerson so hated, to take on opposite or alternative meanings. Business was now a moral term, as in the business of your life. Your commerce was your work in resisting mass culture, what you are told to do. Your profit was your virtue, your principal your principles. Standing outside, being a rooted vagrant was the ultimate goal, as he would write in the closing pages of *Walden:* "I fear chiefly lest my extravagance may not be extravagant enough."

Of course, he still had to make a living, literally speaking, to work off the debt he had got himself into by writing and publishing *A Week,* and he attempted to do so as a lecturer, an attempt that left its mark on *Walden*. After living at the Emersons', Thoreau hit the lecture circuit. Lecturing was something Lidian had encouraged him to do. "You must advertise him to the extent of your power," she wrote to her husband. "A few Lyceum fees would satisfy his moderate wants—to say nothing of the improvement and happiness it would give both him & his fellow creatures if he could utter what is 'most within him'—and be heard." Lecturing was something Thoreau wanted to do too, though he had mixed feelings about it. It didn't pay well, and while he had made it his mission to engage the public since his brother had died, he didn't want to have to lower his artistic bar to please a crowd. But people were almost clamoring to hear about his life in the woods—the trick, if it could be called that, had worked. And he realized that lecturing was valuable as a tool to help him sculpt his writing. Additionally, he had begun to prepare a lecture on his night in jail, the one-night excursion that Emerson had characterized as "mean and skulking and in bad taste." For a while, Thoreau combined lecturing and writing, with some surveying thrown in. After going to Maine, he returned to lecture on it in Concord and the Boston area, then wrote a piece and had it published. He did the same for Cape Cod, as well as Canada, where he apparently had a bad time: In a subsequent piece for the *Atlantic Monthly,* he wrote, "I fear that I have not got much to say about Canada, not having seen much; what I got by going to Canada was a cold." Even if people didn't like the resulting magazine accounts—the Canada trip is on the sour and short-tempered side—they thought he told a good adventure story.

As a lecturer, he was no Emerson, but then few people were—Emerson, as physically awkward as he could be, had the ability to inspire listeners of all backgrounds. But Thoreau was not as bad as is purported by that strain of scholars who chastise him, who describe him, for example, as a "very dull lecturer" and "a bad lecturer," and cite his "always bad delivery." College texts have described him as "a poor speaker and often more interested in the sounds of his own words than the reactions of his audience." But the reviews Thoreau would have read at the time indicate that people thought he could be pretty funny—"constant mirth" is a phrase used after a lecture he gave in Salem, Massachusetts, in 1848. People were interested in seeing the man who had been written up by Horace Greeley in the *New York Tribune,* and they seem to have been almost relieved that he could joke about his experiences at the pond. Especially since he was coming off a commercial failure, Thoreau would likely have been keen to hear a crowd laugh. In this review, from the *Salem Observer,* he is the satirist and Yankee humorist, and the crowd sounds pleased:

> The subject of the lecture was Economy, illustrated by the experiment mentioned.—This was done in an admirable manner, in a strain of exquisite humor, with a strong under current of delicate satire against the follies of the times. Then there were interspersed observations, speculations, and suggestions upon dress, fashions, food, dwellings, furniture, etc., etc., sufficiently queer to keep the audience in almost constant mirth, and sufficiently wise and new to afford many good practical hints and precepts . . . The performance has created "quite a sensation" amongst the Lyceum goers.

Worcester was always a good town for him; they seemed to have loved Thoreau in Worcester, home of his good friend H. G. O. Blake, a Harvard Divinity School graduate who became Thoreau's first disciple. As serious as Blake was, his friends in Worcester tended to think Thoreau was almost hilarious, according to the Worcester *Palladium:*

> This sylvan philosopher, after leaving college (perhaps a little charmed by some "representative" man) [that's code for *Emerson*] betook himself to the woods, where they slope down to the margin of a lakelet of clear water resting upon a fine gravelly bottom. There with a little aid from a brawny Emeralder [that's code for an Irish immigrant], the young man Thoreau erected a house of ample accommodation for himself. Around his house he planted corn, beans, and other esculents, which at a trifling cost furnished him the means of living. At the end of the time he found that he had lived at the expense of about $27 a year, and that his income exceeded his outgoes $13 a year; and that most of his time had

> been given to study, to reading, and to reflection. His lecture was a history of his experience; and is said to have been witty, sarcastic, and amusing.

After a while, Concord was not so good, which makes sense: a person's local reputation is often adversely affected when he or she gains fame outside of home. The symptoms of Thoreau's long-term perceived shortcomings begin in Concord. The locals start to take him down. A man from the town of Lincoln, which is adjacent to Concord, went to hear Thoreau lecture on Cape Cod at the Concord Lyceum in January 1850. "His ideas are strange, many of them, yet I think he had been [*sic*] any other than a 'native' of Concord he would have been well liked by most of the people," the Lincoln resident wrote.

The Thoreau we might imagine from our high school English class—and even from critiques I have more recently read of him—is staid, quiet, offstage. But here this Thoreau managed to "bring down the house," according to reports after his "Economy" lecture in Gloucester. He was always reading draft sections of *Walden* to friends, but he continued to use his lectures as a way of testing material. Simultaneously, in between rewrites, he would return to travel literature, read voraciously in philosophy or natural science; mine the ancient scriptures of India. It speaks to Thoreau's charm that he never met the librarian he couldn't convince to lend him what was supposed to be an unlendable book.

IN ALMOST ALL THE COPIES OF *WALDEN* I HAVE, and I have a few, it is published together with Thoreau's essay "Civil Disobedience." For a long time, I thought of the two works as only somewhat related; they were united, perhaps, by the disdain they both showed Thoreau to have against his town, against civilization. In the popular mythology, "Civil Disobedience" has long been considered further proof that Thoreau was an unsocial crank who reveled in the isolation of jail. But that's wrong, and if you are going to rethink *Walden,* then it's necessary to rethink "Civil Disobedience," to reconcile the two. Far from being a split from society, "Civil Disobedience" is a bold statement of citizenship, of the necessity of putting in your share, and it's part of the formulation of the utopian vision of *Walden*. It is almost *not* about avoiding taxes. Even in skipping some taxes, Thoreau emphasized that there are other taxes he needed to pay, that he was too civil to resist. One of the best lines (or one of my favorites, anyway) has to do with the road. You wouldn't know it in the way it has been given over to the automobile today, but the road is a public space, perhaps the first public space, and when Thoreau talks about the road in "Civil Disobedience," the road is the connection to everything and everyone outside of himself—the road concerns the importance of staying connected. To wit: "I have never declined paying the highway tax, because I am as desirous of being a good neighbor as I am of being a bad subject; and as for supporting schools, I am doing my part to educate my fellow countrymen now."

To consider "Civil Disobedience," I'd like to backtrack to the day in July 1846 that Thoreau landed in jail. The story is tainted in its retelling, but scholars have confirmed the outlines. He was living at the pond. He was in the vicinity of the public square, for whatever purpose. It was three years before the essay would be published. In the street, Thoreau had run into the local constable, Sam Staples. Staples and Thoreau knew each other as citizens of a small town, of course, but also because they were occasional hunting partners. Thoreau had stopped paying his taxes in 1842 or 1843, joining in with the abolitionists, such as Alcott, who did not want to support a government that was against individual liberty. (Thoreau turned up for militia duty in 1844.) When Staples ran into Thoreau in 1846, he told him that he would have to put him in jail if he didn't pay his poll tax. "As well now as any time, Sam," Thoreau is said to have responded. Staples offered to lend Thoreau money to pay the tax; he didn't want to put him in jail. But Thoreau declined the offer. Staples locked him up. By the end of the day, one of Thoreau's relatives, appalled at the notoriety of Thoreau's impoundment, had paid his bail. Thoreau was theoretically free and did not have to spend the night; but as the story goes, Staples already had his boots off for the day—he was in for the evening, in other words, and didn't feel like putting them back on, getting up, and going out again, a personal choice to which civil rights around the world owe a great deal.

Here was a guy who had read all the exploration accounts, both contemporary and ancient, a guy who had decided on explorations at the pond in Concord rather than the Antarctic or the Great Plains, and for him a night in jail was a holiday from his holiday at the pond, as well as a gift to a free-lance writer who used personal happenings as the beginnings of his essays. Thoreau didn't say much about the incident in public until January 1848, at the Lyceum, where he gave a lecture entitled "The Rights and Duties of the Individual in Relation to Government." He subsequently made the lecture into an essay. He was ostensibly resisting payment in protest of the war with Mexico, funded in part by the poll tax, but his night in jail ended up being a much broader action; when the essay was printed, in 1849, the peace treaty with Mexico had been signed for more than a year. Thoreau barely mentioned the war in the article, because in the time that he took to write the essay, he made it about the broader protest against the Mexican War, which was seen as an attempt to expand the legal space of slavery. Slavery is what he was protesting, though not slavery *per se* but the apathy of the people—the well-meaning, the righteous, the speech givers, and the ministers—who purported to be against slavery but in fact fueled a corrupt system. Thoreau was not trying to change a national government that he knew he had little influence over, but hoping to speak to his neighbors. It was addressed to Concord, and to neighbors everywhere, as a personal declaration of independence.

His neighbors were *non*-resisters, and what Thoreau was reacting to, in large part, was the idea of *non*-resistance, the practice of prominent abolitionists at the time. Once again, Thoreau was critiquing a critique. William Lloyd Garrison's New England Non-Resistance Society was formed in 1838. Nonresistance meant refraining from any violence in protesting the government's involvement with slavery, even self-defense, but more than that, it included not cooperating in any way with violence, in this case slavery. Nonresisters would not hold office in a state with a standing army, a police force, or a jail. They saw voting as a kind of cooperation and thus avoided the polls. They did have an escape clause: "We shall . . . obey all the requirements of government, except such as we deem contrary to the commands of the gospel." When Bronson Alcott went to jail for not paying taxes (and lectured on it) in Concord, he called it an act of nonresistance to government. By not resisting, they were peacefully declining cooperation and, thus, separating themselves from the government, doing nothing that supported an institution that supported slavery. This they saw as a separation from the violence of slavery. The nonresisters, as such, are what Thoreau might call "no-government men."

Thoreau's essay can fool you starting out. It begins as if written by a no-government man, along the lines of what we might call a modern Republican. But very quickly Thoreau separated himself from the nonresisters:

> To speak practically and as a citizen, unlike those who call themselves no-government men, I ask for, not at once no government, but at once a better government. Let every man make known what kind of government would command his respect, and that will be one step towards obtaining it.

It needs to be spelled out clearly: Thoreau was *not* a no-government man. He was pro-government, but in a radical new way, or, more accurately, he was applying old lessons in a new way. The essay was originally called "Resistance to Civil Government," an allusion to nonresistance; around the time he died, the title was changed. But as usual with Thoreau, he was thinking about the word *resistance*. For Thoreau the pencil maker, the guy who knows more than we think about machines, resistance refers to a *friction*, a physical force necessary to make a machine work. (It's ironic that today we chide Thoreau for being antitechnology when on this most philosophical point he turns, engineer that he is, to the language and principles of technology) as is often his wont. His complaint with the current government is that most men serve it like automatons: "The mass of men serve the State . . . not as men mainly, but as machines, with their bodies." Men should be serving as men, he argues, with their conscience, their highest sense.

Tax *evaders* might read Thoreau as a hater of taxes, a hater of government. Tax *resisters* might read him, correctly, as a citizen using taxes as they are intended, to power the government; taxes are a citizen's point of purchase. In following his conscience, Thoreau sees a moral obligation to *not* pay. He is working an opposite. To pay is to serve as an automaton. To not pay as a resister is to separate yourself from the system to no effect. To resist by not paying, Thoreau stays within the system. Going to jail is to use the machinery of government itself against itself, a friction.

Thoreau is part of what he calls the "wise minority." Wise minority sounds pedantic, like something more useful at a dinner party in Socrates' time. But the wise minority is unspeakably essential, even now or especially now. Without the wise minority, things can go horribly wrong, in an accumulation of small apathies. It's like when you end up with people who are against torture working in a place like Abu Ghraib, the military prison in Iraq, and being inadvertently involved in torturing people because the system they don't want to disobey condones it, or because the system—and here's the most insidious part—does not condone being *against* the system in any way. The system has no friction! Thoreau might have been a radical anarchist when he merely skipped paying his taxes as an abolitionist, but he turned into something else after he went to jail and wrote his essay. In the time he took to write, he turned the Founding Fathers on their heads, basing independence on the independence of the one, the wise minority. As Lincoln rewrote the idea of freedom with the Gettysburg Address, Thoreau recast his acts as an action not against the government, but *for* it, and he rewrote the obligations of the moral citizen. He became a supercitizen, rescuing the state from the momentum of inaction.

THIS CAN ALL SOUND IMPRACTICAL, but it was executed concretely in the actions of the many people over the years who read "Civil Disobedience" and acted accordingly. The Freedom Riders, riding on buses in Mississippi in 1961, after Congress had ruled against segregation in interstate transit, did not attempt to destroy the government by being arrested for breaking state segregation laws; their actions ultimately improved it. In riding the buses in order to be arrested, they used the system against itself, a friction. When they filled up the jails in Mississippi and refused bail, they were the frictions that helped make the system work on behalf of the rest of us.

Thoreau wasn't ready for the state to disappear. He wanted to recalibrate his own relationship with it. This is what Martin Luther King, Jr., liked—particularly social change through tax resistance, a feat that links Thoreau to the Algonquins who refused to pay taxes to the Dutch in New Amsterdam, and to the American revolutionaries (whose tax resistance was a more modern example from the vantage point of 1847) when they refused to pay the Stamp Tax in the 1760s, as an act meant to restore their rights as British citizens. Thoreau managed to meld the Christian pacifism of the Quakers with the political liberalism that powered the revolution of 1775.

> All men recognize the right of revolution; that is, the right to refuse allegiance to, and to resist, the government, when its tyranny or its inefficiency are great and unendurable. But

> almost all say that such is not the case now. But such was the case, they think, in the Revolution of '75. If one were to tell me that this was a bad government because it taxed certain foreign commodities brought to its ports, it is most probable that I should not make an ado about it, for I can do without them.

He may be tough to get along with, but he wants us to get along. He was by no means a fan of politics; to a large degree, he shares the other Transcendentalists' disdain for the masses, and he is not about to run for governor. He sees that he is part of his government, even if he sometimes chooses not to be, that there is no way to abandon the ship of state, for it sails with you whether you buy a ticket or not.*

Thoreau does not debate the war on Mexico in the essay. It is unjust—that's a given for him. The government, in executing the war, relies on the inaction of the majority of men. Thoreau does not dwell on the stream of funding that allows the war; the poll tax that Thoreau was not paying did not actually pay for the federal government's invasion of Mexico, a fact that irked Emerson, who called Thoreau's time in jail misguided. The federal government used other taxes for that. But here, Thoreau pointed out something that

* He takes the idea of friction a step further, where it gets complicated, to say the least: What if a friction became itself a machine? In a United States where slavery is part of the mechanics of society, Thoreau presents slavery as a friction that is a machine unto itself:

> All machines have their friction; and possibly this does enough good to counter-balance the evil. At any rate, it is a great evil to make a stir about it. But when the friction comes to have its machine, and oppression and robbery are organized, I say, let us not have such a machine any longer.

Slavery has to go in its entirety, in other words, as a system.

we have all realized by now—that the government*s*—local, state, federal—had blended into one government, and that the easiest, the most *civil* (a pun that Thoreau would have intended) manner in which to meet the federal government is in the person of his neighbor, Mr. Sam Staples, constable:

> I meet this American government, or its representative, the State government, directly, and face to face, once a year—no more—in the person of its tax-gatherer; this is the only mode in which a man situated as I am necessarily meets it; and it then says distinctly, Recognize me; and the simplest, the most effectual, and, in the present posture of affairs, the indispensablest mode of treating with it on this head, of expressing your little satisfaction with and love for it, is to deny it then.*

To deny the tax collector is the simplest way to revolutionize, the practical way—so said the practical philosopher who was also a local handyman. It was an action, most importantly, whereas the abolitionists were, from his viewpoint, mostly involved in talk. (It's the difference too between Emerson and Thoreau, at this point in their relationship.) At the dawn of the era of machine politics, Thoreau operated on the level of one man. He believed in the power of the loyal opposition.

> I know this well, that if one thousand, if one hundred, if ten men whom I could name—if ten honest men only—ay,

*Staples, the jailer, was also a state representative about to be up for election, and the majority of Concord was *for* the Mexican War; so jailing a protester might have been an electoral benefit to him—just one possible reason he and Thoreau stayed friends. He would have appeared tough on war resisters.

> if one HONEST man, in this State of Massachusetts, ceasing to hold slaves, were actually to withdraw from this co-partnership, and be locked up in the county jail therefor, it would be the abolition of slavery in America. For it matters not how small the beginning may seem to be: what is once well done is done forever. But we love better to talk about it: that we say is our mission. Reform keeps many scores of newspapers in its service, but not one man.

The mere existence of taxes emphasizes the fact that you can't change the system all at once, all on your own, he says. You can't sit down at all the desks of government and run the various shows. You can't—as in, you are not able—and you can't in that you have a life to live, things to do, like write a book or take care of the kids (in Thoreau's case, somebody else's). But you have the moral obligation to somehow separate yourself from screwing up other people's lives as best you can.

> It is not a man's duty, as a matter of course, to devote himself to the eradication of any, even the most enormous, wrong; he may still properly have other concerns to engage him; but it is his duty, at least, to wash his hands of it, and, if he gives it no thought longer, not to give it practically his support. If I devote myself to other pursuits and contemplations, I must first see, at least, that I do not pursue them sitting upon another man's shoulders. I must get off him first, that he may pursue his contemplations too.

Life is complicated, and it is full of minor inconsistencies, little frictions, but when your entire town, or nation, becomes an inconsistency, then things are, frankly, gross.

And when Thoreau talks about gross inconsistencies, he's talking about people who are against slavery but, well, don't want to ruin the economy. He's likewise talking about people who were buying grapes in the 1970s and saying they supported César Chávez and the United Farm Workers. He's talking about people who are against an oil war but using more rather than less of the gas that drives it. He's talking about all of us, and the inconsistencies we try to avoid, or put aside, or forget about.

The *Concord Freeman* (which Thoreau decried, but read) was, for example, against the war, but supported funds for the troops, not wanting to seem to be against the well-being of the troops. The illogicalness of this is what burned Thoreau:

> See what gross inconsistency is tolerated. I have heard some of my townsmen say, "I should like to have them order me out to help put down an insurrection of the slaves, or to march to Mexico—see if I would go"; and yet these very men have each, directly by their allegiance, and so indirectly, at least, by their money, furnished a substitute. The soldier is applauded who refuses to serve in an unjust war by those who do not refuse to sustain the unjust government which makes the war; is applauded by those whose own act and authority he disregards and sets at naught; as if the state were penitent to that degree that it hired one to scourge it while it sinned, but not to that degree that it left off sinning for a moment.

Contemporary translations are various, but suffice it to say he is commenting on the person who buys organic food and has it delivered in a big polluting truck that sits idling in front of his or her house or the developer who builds luxury eco-housing at the expense of the low-income housing market. The stereotypical Thoreau seems to call himself privileged in his separate and monastic life, but the real Thoreau was actually more like us, with a life to live, things to do, inconsistencies to deal with, one person in the flock of citizenry. He is not even a radical radical, but a citizen made radical by his circumstances. He is *forced* to act. Rosa Parks does not go out of her way to battle racial injustice. She rides on a bus to work, *her* bus, *her* route. The tax collector forces himself on Thoreau, who is forced to make a decision between paying taxes to support a slave state and not paying taxes, which results in jail: "Under a government which imprisons unjustly, the true place for a just man is also a prison."

In the end, binding *Walden* together with "Civil Disobedience" seems like a good idea to me. "Civil Disobedience" is part of *Walden*. It's the owner's manual. People look to *Walden* for direction, when it is more importantly inspiration. "Civil Disobedience" tells you how to act, but the prescription is not as simple as "move to the woods and plant a garden." The writer of *Walden* is trying to imagine the place where the civil disobedient self would live, a community of potential government resisters, and how that would go.

AS HE PROCEEDED THROUGH DRAFTS OF *WALDEN*, and after his second and last stint at Emerson's house, Thoreau moved back home, into the attic of a new home called the Yellow House that the family had purchased in the center village. The Thoreaus traded up. After all their hard work, the Thoreaus had finally made it. Thoreau was revising *Walden* again and again, often layering into it his latest readings, notes, and quotes and descriptions from his journal.

Meanwhile, he was elected a member of the Boston Society of Natural History. On his daily and regular walks in Concord and its environs, he collected plants and animals for Harvard, sometimes storing samples in a compartment he'd made in the top of his hat. This was odd, but maybe not quite as odd as it sounds. Collecting plants was yet another trend at the time, though the hat was his own idea. He went to Cape Cod on a long trip, using it as material for lectures and then for a series of articles; when he left town—to lecture in New Bedford or Worcester, for example—he often carried his extra clothing in a brown paper bag, or in a bag on a stick, a hobo. He turned his Canada lecture into a piece that he sent to Horace Greeley, who said it was way too long. He designed a pipe-bending machine for George Loring, a lead-pipe maker in Concord, drawing out a working diagram for Loring. (He frequently used his drafting and engineering skills on his neighbors' behalf, designing everything from a piston and cow stanchion, for milking, to a device that allowed him to build a fire on the front end of a boat, to illuminate the bottom of a pond or river while rowing friends.) He sold excerpts of *Walden* to magazines, though in the end he didn't get paid for them. In the summer, he went swimming, as was his wont, often with Channing, who was shocked when Thoreau showed no embarrassment upon the arrival of a hunter, the swimmers being unclothed.

In between the publication of *A Week* and the publication of *Walden*, Thoreau changed the way he approached journal writing, and it was to be the most significant change in his life as a writer, and thus perhaps in his life. His journal entries became even more regular, more wide-ranging and soaked more thoroughly in local lore and fact. The journal became a kind of daily newspaper of his observations; it was less a collection of short essays (though those showed up on occasion) and more straight reporting, at certain times he seemed like a court reporter roaming a New England town. He lived in his parents' house, and every day walked for four or six hours and wrote several thousand words, starting with notes in pencil in the field, which were transferred and expanded on, using ink, in his journal.

At first, he would tear pages from his journal and work them directly into magazine stories. Now, the journal was a thing unto itself. Ideas might be worked up in the journal, but the journal was a work of its own. He continued with his surveying—in the summer of 1850 he laid out the route for a road from the railroad station in Concord to the Mill Dam. Sometimes he observed for the sake of surveying, sometimes the sake of his journal, oftentimes both. He worked for real estate developers, helping lay out a road, and he took notes in the field and revised and copied them over when he got home, making more and more lists of birds and plants, noting the dates of leaves turning color and falling, noting the flowering of plants, for himself. He took down everything, from the dates houses were built to overheard local conversation. Ironically, it is more of the kind of journal that we often think of Thoreau offering in *Walden,* and to be sure, many observations that first appeared in the journal found their way into *Walden*.

But as he pressed on with *Walden,* he began to condense the two-year time line into one, and rather than a strict account of a year, he designed a symbolic one. His journal documents his life at Walden and subsequently in Concord. As he read Vedic myths, as he punned on New Testament scriptures, *Walden* was becoming a myth more than a story, a joke that developed more and more serious and sometimes spiritual punch lines, written to inspire modern citizens to break out from the lockstep of culture and in so doing make a new connection to their community. A few days before he signed with a publisher, he bought a telescope in Boston. On March 15, he painted his boat. On March 16, he signed with Ticknor and Co., and in the summer of 1854, he received his first copies of *Walden*.

A LOT OF *WALDEN* IS HARD GOING, for this reader, anyway. The relatively easy setup of the first chapters leads to some places where the woods are dark and deep, so that sometimes I want to throw it down. But then the imagery is of such beauty—"It was not lonely," he writes of a hawk circling in the sky, "but made all the earth lonely beneath it"—that you linger, and as you attack the layers, you begin to see the complexities, and then the greater beauty in the complexities. Robert Frost said a perfect poem runs itself and carries away the poet with it. "Read it a hundred times and it will forever keep its freshness as a metal keeps a fragrance," and so it is of *Walden*. *Walden* is the light on at night for the person on the back road, for the tired traveler wondering how much farther they have to go and then realizing that the path is what matters. In *Walden,* Thoreau sits peacefully in his cabin doorway, rising when the world believed him to be rotting: "I grew in those seasons like corn in the night."

Walden takes the long way around on purpose, making it in itself representative of Thoreau's life. With the book, he was not suggesting everyone live as he did at the pond, or as he ever did in Concord: "I would not have any one adopt MY mode of living on any account." To repeat: if you see *Walden* as a manual, which it was not intended to be, then you will build a house in the woods and try to live a solitary life, but if you see it as an edifice in itself, you can enter and allow it to engage you. Better to see it as a mental striptease: with language that strips you down, peels away conceptions, and prepares you for a new interpretation, to be an extra vagrant, a civil disobedient, an independent resister. You can choose your reading, of course, but given that Thoreau has gotten so caught up in our ideals of nature, that he is thought of chiefly as a defender of the rural, the natural—that is, rural life—it seems important to describe the ways in which the experiment at Walden Pond can be seen as a view of a civilized or even urban world.

In my own reading, I was startled to find the city, of all places, in *Walden*. I initially went to Walden thinking of it as the archetype of rural living; I was fishing for Thoreau's antiurban musings, as I mentioned, looking for the beginning of the great divide between us and nature, between the man-made and the non-man-made, nature and the city. But wading in, I discovered *Walden* offered something else entirely. I found a vision of a city, a new metropolitan plan. The most complicated part of it all—and at the same time the simplest—is that the city envisioned in *Walden* is not actual, and that *Walden* is not about building a house or a place but rather about building a life and community, starting with you and your own.

It's hokey, I know, but try imagining that the narrator of *Walden* is an architect. Think of him less as an architect of your vacation home and more like an architect of your town or city or life, as someone who is working on a new way for all of us to live. He's a planner more along the lines of Buckminster Fuller, a scientist and artist who sees great possibility in the rearrangement of our living situation, who is not against technology but frustrated by misapplication, it's *un*practical application.* Picture the architect of *Walden* reviewing the needs of his client, who in this case is us. "The mass of men lead lives of quiet desperation"—this is the situation that the architect hopes to relieve, the problem his "building" will hope to solve. Society seemed to have no center; the old village king was dead; numbers ruled the day as the sole calculations of enterprise. Our architect sees where we are living and how we are living, and he sees, as he might mention to a TV morning-show host, that there is bad *feng shui*. He notes in his presentation to us that the reform-minded citizens of the community worry about slavery but do not recognize that they are themselves enslaved. "It is hard to have a Southern overseer; it is worse to have a Northern one; but worst of all when you are the slave-driver of yourself." He points to a local example, the teamster, riding his horses, more worried about the horses than himself. "What is his destiny to him compared with the shipping interests?"

So the architect chooses a spot, a place to site a new life. This is what is complicated about *Walden,* what leads to so much misinterpretation. *Walden* was doing just what Thoreau's friends at Brook Farm were doing, only theoretically. *Walden* is the utopian community for your imagination, and

* Buckminster Fuller shares a birthday with Thoreau, and a philosophical cast. The grandnephew of Margaret Fuller, Fuller was a scientist who loved art, who distrusted pure fact, and was cranky in the same way as Thoreau, not so much because of pessimism as because of optimism. (Why is it so difficult for people to imagine cranky optimists?) In 1964, *Time* profiled Buckminster Fuller: "He is a throwback to the classic American individualist, a mold which produced Thomas Edison and Thoreau—men with the fresh eye that sees and questions everything anew, and the crotchety mind that refuses to believe there is anything that cannot be done. What Fuller sees excites him with the vision of man's potentialities, and he has made it his mission to help man realize them. Says he: 'Man knows so much and does so little.' "

where the real utopias have failed, it succeeds, and not just because it has no membership fees, no manure-related work to divvy up. It succeeds because it takes you to a place where you might jump-start your spirit. It is the example of retirement—from the rat race. And as a work of art, as a book, it is proof, in itself, that it is possible: the experiment at Walden yielded *Walden;* you are holding the evidence in your hands. In a world that was preventing people from doing much more than keeping up, inspiration was a central goal for Thoreau rather than a side benefit. Not that everyone needs it—Thoreau is always clear in saying that some people are just as well off (or even better off) without him. Most people, though, are just stuck, stuck in a world that tells them to be better but offers empty solutions, a world that runs real reform just ahead of them like the rabbit at a dog-racing track.

> I do not mean to prescribe rules to strong and valiant natures, who will mind their own affairs whether in heaven or hell, and perchance build more magnificently and spend more lavishly than the richest, without ever impoverishing themselves, not knowing how they live—if, indeed, there are any such, as has been dreamed; nor to those who find their encouragement and inspiration in precisely the present condition of things, and cherish it with the fondness and enthusiasm of lovers—and, to some extent, I reckon myself in this number; I do not speak to those who are well employed, in whatever circumstances, and they know whether they are well employed or not;—but mainly to the mass of men who are discontented, and idly complaining of the hardness of their lot or of the times, when they might improve them. There are some who complain most energetically and inconsolably of any, because they are, as they say, doing

> their duty. I also have in my mind that seemingly wealthy, but most terribly impoverished class of all, who have accumulated dross, but know not how to use it, or get rid of it, and thus have forged their own golden or silver fetters.

In siting his community, Thoreau uses terms that sound familiar, that you, the client, will understand. The architect describes a place where there is a good "port" (the pond). He cites the proximity of the rail line (the nearby Fitchburg railroad that cut across the edge of Walden). These are connectors, of course, lines out and in, not the infrastructure of hermitude. There will be capital invested in the project, the architect notes, but the capital will be you. The architect in *Walden*—Thoreau's "I," that is—is part huckster, a somewhat refined version of P. T. Barnum, and at the outset of the book, *Walden*'s architect has the promoter's audacity to compare it favorably to another city, the great Saint Petersburg:

> I have thought that Walden Pond would be a good place for business, not solely on account of the railroad and the ice trade; it offers advantages which it may not be good policy to divulge; it is a good port and a good foundation. No Neva marshes to be filled; though you must everywhere build on piles of your own driving. It is said that a flood-tide, with a westerly wind, and ice in the Neva, would sweep St. Petersburg from the face of the earth.

"PILES OF YOUR OWN DRIVING" is another way of saying that you won't find a pattern book from which to cut *this* good life. You won't read about it in Martha Stewart's *Living* or in the Home section of the *New York Times,* because

Walden is itself a true enterprise, a free economy that is truly free, where trade and commerce propel the place rather than destroy it—it's a state of mind, in other words, that leads to a state of being, the good (or better, anyway) life. It starts with you and builds out. Recall that everywhere in Concord, farmers are working themselves ferociously, ostensibly for themselves, but in fact for the banks. From *Walden:*

> On applying to the assessors, I am surprised to learn that they cannot at once name a dozen in the town who own their farms free and clear. If you would know the history of these homesteads, inquire at the bank where they are mortgaged. The man who has actually paid for his farm with labor on it is so rare that every neighbor can point to him.

Walden imagines a way of living where farms are owned by farmers, where houses are owned by the people they house, a home life that is mortgage-free. This is romantic as it is, even today, radically optimistic. Thoreau parodies the Gospel of Matthew to make a point that separates his envisioned home from the second homes of the retiring wealthy, a swipe at a lot of people all at once. In Matthew, Jesus says, "Foxes have holes and the birds of the air have nests, but the Son of Man has nowhere to lay his head." In *Walden*, the narrator says, "I think that I speak within bounds when I say that, though the birds of the air have their nests, and the foxes their holes, and the savages their wigwams, in modern civilized society not more than one half the families own a shelter."* Resources are mismanaged. In fact, *Walden*'s architect imagines a house even smaller than the house that Thoreau built, a box, six feet by three feet, that he has seen Irish laborers store their tools in. He jokes that it might be augmented with breathing holes and be suitable for housing. "You could sit up as late as you pleased, and, whenever you got up, go abroad without any landlord or house-lord dogging you for rent," he says in *Walden*. "Many a man is harassed to death to pay the rent of a larger and more luxurious box who would not have frozen to death in such a box as this."

The architect has designed a city owned by its citizenry, as opposed to what a town like Concord was becoming: an association of mortgage holders who, moreover, were beginning to think of themselves mostly as consumers. It sounded as if he was joking about living in a box, and he was, but then again he wasn't—a problem Thoreau knows he has. "I am far from jesting," *Walden* goes on. "Economy is a subject which admits of being treated with levity, but it cannot so be disposed of." Thoreau thought of the economy of living as, in his words, "synonymous with philosophy." Economy comes

* *Walden* continues: "In the large towns and cities, where civilization especially prevails, the number of those who own a shelter is a very small fraction of the whole." People today may read *Walden* for insights into organic gardening, which is great—*Walden* has a lot to offer. But as I type, a lending crisis has the number of people who own homes decreasing: nearly one in ten homeowners faces foreclosure in late 2008, an example of *Walden* becoming *more* true. On the one hand, Thoreau offers a tonic: your home will no longer own you. On the other hand, when you read an environmental magazine suggesting re-engineering the world through the use of luxurious low-water-usage kitchen products or by planting self-sustaining gardens on your property, you can think, what about the people who don't own kitchens or property? Almost seventy percent of the country owns their home (a number that is likely to decrease significantly by the time this book is published). The rest doesn't.

from the Greek word *oikonomia,* which means *household management.*

The architect then offers references to the past, not for the sake of nostalgia, but for proof that the new project is possible. He cites *Wonder-Working Providence,* by Edward Johnson, who, in 1653, described the settlement of New England—farmers first building cellars, then covering them, then associating with neighbors. In Johnson's time, land was held in common, ensuring the continued existence of farms, ensuring property dispersed by need and intrafamily association with interdependent interests. The architect of *Walden* was not against capital; he was only wary of the way it flowed. He deplored investment *away* from town, into the banks that owned the farms the farmers farmed. His new site, being on Emerson's property, was itself in the ring of farms that formed Concord—the ring of failing farms, by the way. From the vantage of Walden Pond, even Concord, however, was urbanizing. It was not Boston, but it was more and more like Boston. The day was organized around the arrival of the train. Tenantless farmers worked land they did not own, the woods were filled with work-searching migrants. By some definitions, a slum represents the fabric of civic life coming apart, and this is where *Walden* is set, where the landless squat. (The word *slum* first appears in English around this time, and has most recently been seen as deriving from Irish Gaelic: *'S lom,* pronounced *s'lum,* means *is bare,* or *is naked, is poor*—that is, a vulnerable place.)

Work was wasted, both the backbreaking and the mental kinds. This is *Walden*'s central critique, as Thoreau sets out to design a more practical way to live. Nature proved the point, even whimsically: the farmers work for the banks and the bank managers, but the squirrels work for themselves. In homes around Concord, parlors were designed specifically for conversation, but Thoreau hears much conversation that is divorced from the work of the home and work of life in general. Thus *Walden* points out the holes in our complacency: "It would seem as if the very language of our parlors would lose all its nerve and degenerate into palaver [nonsense] wholly, our lives pass at such remoteness from its symbols, and its metaphors and tropes are necessarily so far fetched, through slides and dumbwaiters, as it were; in other words, the parlor is so far from the kitchen and workshop. The dinner even is only the parable of a dinner, commonly." Can we today imagine beautiful dining rooms where people rarely eat and have nothing to say? In the new place that *Walden* describes, the parlor is more like a public square than a parlor. Conversation is general, as in nature, as it might be, as he sees when he scrapes away snow from the pond:

> I cut my way first through a foot of snow, and then a foot of ice, and open a window under my feet, where, kneeling to drink, I look down into the quiet parlor of the fishes, pervaded by a softened light as through a window of ground glass, with its bright sanded floor the same as in summer; there a perennial waveless serenity reigns as in the amber twilight sky, corresponding to the cool and even temperament of the inhabitants. Heaven is under our feet as well as over our heads.

I would be a much better writer if I could resist a pun myself, if I could avoid saying that the inspiration for a new life is not just at hand but also at foot.

THE NARRATOR IN *WALDEN* WEARS A SECOND MASK in the second half of the book. He begins as the self-deprecating Yankee smarter than he lets on: "I never knew, and never shall know, a worse man than myself." In the America obsessed with success, Thoreau stars as bum or hobo, the reporter to a journal "of no very wide circulation." In the second stage of the book, centered around "Higher Laws," he becomes the smart guy who is impossibly smart, who seeks to live among the stars, even though the stars are purely for navigational purposes, a GPS to help you find your own path through actual life on the ground. Thoreau knew people thought of him as the former, and he played with it. He also knew that neither persona quite worked. If he were setting up an "experiment" today, Thoreau might joke in his first half about driving an alternative-fuel vehicle with two wheels and pedals to a "luxury" eco-vacation-home that was a tent in a field. He might squeeze an orange with five fingers and a knife in lieu of a $2,000 juicer. In the second half, he might ponder the nature of the universe and morality as it seems to a twenty-first-century human. These two narrators of *Walden* will come back to town at the end of the book as one. For my purposes, I'd like to dwell on the fact that the idea of town and country have simultaneously changed places as well. This is a big trick in *Walden*. As far as living goes, Thoreau has turned nature around.

Around the time of the Puritans, America had a theoretical problem with nature; all that could not be farmed or utilized or harnessed in one way or another was worthless—"waste," the Puritans called it. Nature, a fallen world, had been banished from the town, the only place the saved could hope to prosper, or at least work to keep themselves saved. The Transcendentalists had turned against the Puritans and gone back to nature, declaring it a source of an inspiration that was necessary to balance out the crass materialism of American life. For Thoreau nature was the path to the divine, not an end in itself.

This is the part that our popular conception of *Walden* misses. Thoreau's nature book was no lark in the woods. It was a book about reviving the institutions of the metropolis, reawakening them the way leaves return to branches in the spring and, in so doing, restore the forest. Thoreau was using the pastoral ideals to reinfuse eighteenth-century urban ones: the port city, with its connections to the outside world; the republic, each noble citizen charged with virtue and higher learning, where Jefferson's farmer was a thinker and a voter; Virgil composing a second *Georgics;* technology worthy of the tasks it streamlined. The place being established at Walden was a city, full of people who had managed to reform themselves by stepping outside for a minute or an hour or a couple of weeks or years. It was a community of inspired individuals. The "I" of *Walden* becomes a "we," an inspired we. The economics guy is one with the philosopher. "Truly we are deep thinkers, we are ambitious spirits!" he exclaims. By the conclusion of the book, the "I" becomes "you." But like a chrysalis, the "you" is a new you:

> Who knows what beautiful and winged life, whose egg has been buried for ages under many concentric layers of woodenness in the dead dry life of society, deposited at first in the alburnum of the green and living tree, which has been gradually converted into the semblance of its well-seasoned tomb—heard perchance gnawing out now for years by the astonished family of man, as they sat round the festive

> board—may unexpectedly come forth from amidst society's most trivial and handselled furniture, to enjoy its perfect summer life at last!

The dead wood of society has life in it after all, and *Walden* meant to make you stop and wonder about whether you are dead wood or the live bug, if you are waiting quietly or coming alive.

In *Walden,* Thoreau imagines a place that is cosmopolitan, and contrasts it with the provincial Concord—no wonder so many of his neighbors would always hate him! If he can be faulted, he can't be faulted for the things he is usually faulted for, like looking backward. He can be faulted for being too zealous about the future, too optimistic, too certain, perhaps, that so much of what modernity offers will benefit a place—"If we live in the Nineteenth Century," Thoreau asks, "why should we not enjoy the advantages which the Nineteenth Century offers? Why should our life be in any respect provincial?" He did not say, romantically, that we ought to revert to the institutions of the past. He is not the person he describes as "the village do-not," the one who is satisfied with the way things have always been: "Let the village do-not stop short at a pedagogue, a parson, a sexton, a parish library, and three selectmen, because our Pilgrim forefathers got through a cold winter once on a bleak rock with these." *Walden* has higher aspirations. Every inhabitant is to be like a nobleman, as opposed to an overworked dog or a farmer of debts, and everyone involved will aspire to associate completely with everybody else, in the revolutionary way laid out in the founding documents of the United States—to cultivate a kind of person, in fact, that is maybe *better* than a nobleman:

> As the nobleman of cultivated taste surrounds himself with whatever conduces to his culture—genius—learning—wit—books—paintings—statuary—music—philosophical instruments, and the like; so to act collectively is according to the spirit of our institutions; and I am confident that, as our circumstances are more flourishing, our means are greater than the nobleman's.

LIKE I SAID, I PERSONALLY WENT TO *WALDEN*, the book, to find a Thoreau who was against associations and associating, who was all about going solo; I was writing a book about cities and proposing that people naturally tended toward groups. To my surprise, I found a champion of cities in Thoreau. *Act collectively!* he said. Stick together! Join the club and pay the dues, and don't abandon the ship, even if you have to get arrested and thrown into the brig to save it, even if you feel undervalued. Technological innovations are fine as long as they are put to good practical use. Take the newly invented steamer to England but have a reason to be there. Send a telegraph message but have something to say. Use text messaging, but for more than delivering the news that, as the narrator of Walden jokes, ' "Princess Adelaide has the whooping cough.'

"After all," Thoreau goes on to say, "the man whose horse trots a mile in a minute does not carry the most important messages; he is not an evangelist, nor does he come round eating locusts and wild honey." The architect is touting quality, a quality based on the quality of the citizen, of the noble villages:

> New England can hire all the wise men in the world to

> come and teach her, and board them round the while, and not be provincial at all. That is the uncommon school we want. Instead of noblemen, let us have noble villages of men. If it is necessary, omit one bridge over the river, go round a little there, and throw one arch at least over the darker gulf of ignorance which surrounds us.

Just typing this book, I have, as mentioned, aged, but the older I get the more it becomes clear that the long way is probably the better way. You don't have to be Bill Gates, who just retired and is ready to give away all the money he spent decades accumulating, to know that this argument worked then as it does now: we are racing somewhere, but we're not exactly sure about where. The way has to present itself. Just in the way he made it, *Walden* is the example of the long way around: written while writing another book, then, after *A Week*'s failure, written over and amended for years and years. And despite the people who flock to the pond to somehow see the pond as described, *Walden* is a place that Thoreau intended for you to make on your own—not an imaginary place, but a real place for the imagination. And yet we want to match the "I" to the Thoreau.

I know I do. How can you not? Thoreau the person is like a bur on your sweater that you pick up after walking through the woods. The story from Thoreau's life that I have attached in my own mind to the end of *Walden* has to do with some seeds that he found in the basement of an old house. After *Walden* was completed, he heard that a nearby house built before the Revolution was being demolished. Thoreau loved ruins, a way to tour the past on foot. When he got there, only the foundation remained. In the basement, he discovered some seeds—the seeds of a plant that no longer grew in Concord. He planted the seeds. They germinated. The plant returned to Concord. It's a little bit like a science fiction story that would involve dinosaur DNA, and it's a little like the end of *Walden,* the bug that hibernated in the tree hatching years later from the leg of a table. Thoreau's seed recovery is a resurrection story, but it's a very practical resurrection story, brought on not by any external power, but by internal facilities, a gift within. Thoreau was just paying attention, that's all. Just looking at what was going on around him.

THERE'S OBVIOUSLY A LOT MORE TO *WALDEN* than what I have been going on about at length here. There are puns scattered everywhere, like pine needles on a forest floor, and arguments are woven slowly like baskets, and generally, the more you go back to *Walden,* the more sounds you hear, little reverberating drips on the water's edge, the echoes of Thoreau's flute as he sits in his boat and plays, sounding the depths of Concord and its environs. Even the idea of hermit was a trope that Thoreau was playing with—antebellum newspapers were often claiming to have found a soul in a cave, a man in the mountains. Thoreau hoped the spiritual associations of hermitage that he planted in *Walden* would grow and thrive, but in real time the author of *Walden* kept a chair by his door to encourage visitors, and took it away only when he was writing, which, as I shift from interruption to interruption in the course of this typing, I realize is at the very least a writer's utopia, actually.

But there's one point I'd like to emphasize before getting on with Thoreau's last writing and the end of his life. In this so-called nature book, there is no division between what we today commonly think of as natural and nonnatural. In fact, he seems to attempt the opposite, especially in the famous (to Thoreau scholars, at least) sand-cut section of *Walden*.

The sand-cut passage is a lot of work, if you ask me. It is Thoreau's description of a frozen bank of sand, the sand beneath the railroad track, as it melts in the late winter sun, and it reads more like *Finnegans Wake* than *Ulysses*. The sand-cut section has been ferociously studied, and I might earn a PhD just tearing apart the first paragraph, so, degreeless, I do not intend to break new ground. For a long time, the sand-cut section was considered mysterious, its secrets lost, but over the last few years, it has come to be seen in relation to Thoreau's interest in philology and wordplay, in the light of how seriously he took his jokes and puns.

The sandbank cut through the woods, the path of the railroad track between Thoreau's house at Walden and town—he often walked along the railroad cut to go to and from Concord. In other words, the sand cut was the very line between what we would call natural and man-made. If I were Thoreau and I were looking for a similar area in today's world, I could still go to a train-track bank, or I might go to the edge of a highway, a spot where not too many people would be, a borderland, an edge. It's one of those places that springs up almost accidentally when we make a passageway and cut through one place to get to another. "Few phenomena gave me more delight than to observe the forms which thawing sand and clay assume in flowing down the sides of a deep cut on the railroad through which I passed on my way to the village, a phenomenon not very common on so large a scale, though the number of freshly exposed banks of the right material must have been greatly multiplied since railroads were invented," he wrote.

To get at what he was trying to do, recall that he believed words were rooted to the original ideas of things, which in turn meant they were linked to the divine. Words were like a connection to God's mind, via several extension cords; or, to put it another way, we are hardwired to Creation. We are so hardwired, in fact, that just in saying certain words, we get to experience aspects of life and death, as if we were ourselves a video game being controlled by the Logos. "When the frost comes out in the spring, and even in a thawing day in the winter, the sand begins to flow down the slopes like lava, sometimes bursting out through the snow and overflowing it where no sand was to be seen before," Thoreau writes. He describes the process, the shapes of the flows, the patterns of the thawing, the melting sands, using the words that are mimicking creation. It gets crazy-sounding—"In globe, *glb,* the guttural *g* adds to the meaning the capacity of the throat"—and understand that the language theories that Thoreau was working with have subsequently been debunked and reworked. Essentially, though, Thoreau is showing the relationship between the art of language and the creation of life, showing the force of life is everywhere, even out by the railroad bank, even in what we would today see as an abandoned lot or industrial area, a highway strip, out-of-bounds.

"What is man but a mass of thawing clay?" Thoreau wrote. It may be the peak of his book, the climax, when he stands as an artist witnessing the creation of nature, and it is not in a pristine meadow or a virgin forest, but on the equivalent of the side of a busy road. Imagine Thoreau standing there marveling today, until someone called the cops, or he got mugged, or hit by a car. (When the cops came, they would find his person covered with scraps of paper filled with pencil scribblings, with plant samples, leaves, bugs, plant and animal whatnots, in the top of his hat.)

And what does the shape of primordial life look like to a man who meticulously describes all that is around him? What do these shapes resemble to Thoreau the country bumpkin? Not a tree or a fish or a bird, but, as he said in the passage, "towns and cities." At the ground zero of life's beginnings, in the oozing, volcanic moment of first spring, our premier nature writer saw the city.

Chapter 10

AFTER *WALDEN*

BOOKS ARE A GAMBLE, a fact authors come to accept, eventually. One book flies or maybe soars and another sinks like a stone, the conditions for a healthy birth depending on too many forces to factor. If Thoreau seemed more at ease on the publication of his second book, perhaps it was because he knew that bookmaking is like planting beans, you can do everything right and still be wiped out by a storm or a frost. The commercial disappointment of *A Week on the Concord and Merrimack Rivers* prepared him for the publication of *Walden*. At this point a little more than half-way through his time in publishing, he was both more hopeful and less hopeful. Just before it appeared in book-stores, Thoreau mused to himself on his faults—there is nothing like second-guessing what you have written, as this author can attest to you at this very moment:

> My faults are:—
> Paradoxes,—saying just the opposite,—a style which may be imitated.
> Ingenious.
> Playing with words,—getting the laugh,—not always simple, strong, and broad.
> Using current phrases and maxims, when I should speak for myself.
> Not always earnest.
> "In short," "in fact," "alas!" etc.
> Want of conciseness.

And yet his hope was like his descriptions of spring in his journal, always returning, each time eliciting an original excitement—he's thought of as a curmudgeon, but perhaps Thoreau's greatest skill was recognizing and cultivating happiness, keeping it burning, even in the smallest ember, the way Northwest Coast Indians carried fire in boats on long ocean trips. But by the time of publication, in the summer of 1854, he was, as Emerson described him, "in a tremble of great expectation." What he thought he really ought to do, once again, was settle down and buy a farm.

Using all his connections, pushing hard for a good bargain, he got a better than average advance for *Walden,* as well as a brown cover and two illustrations—a map of the pond by Thoreau, and a sketch of the cabin by his sister. Initially, it sold well, at least in comparison to *A Week.* Good reviews rolled in slowly. The *Providence Journal:* "shrewd and eccentric." The *Christian Register:* "We suppose its author does not reverence many things which we reverence; but this fact has not prevented our seeing that he has a reverential, tender and devout spirit at bottom. Rarely have we enjoyed a book more, or been more grateful for many and rich suggestions." Bad reviews also appeared, in which the seeds for the Thoreau everyone knows were planted: "It is difficult to understand that a mother had ever clasped this hermit to her bosom, that a sister had ever imprinted on his lips a tender kiss." The *New York Times* called it a comedy and did not mean that as a compliment.

He had sneaked weaponry—philosophy, social criticism, a call to prayer even—into an ordinarily bucolic and innocuous sort of writing that people usually went to for diversion: nature poetry. As a result he was attacked, oftentimes for things he was not supporting at all. He was attacked, for instance, for cynicism, pantheism, unsociability. He was praised for originality and humor, a critical mix that seemed to make Thoreau happy on balance. He sent books to friends, and received letters of congratulation. Hawthorne in particular liked it, and noted that Thoreau would most likely pretend praise was as nothing to him, even though, Thoreau being human, it meant a lot. Hawthorne wrote to a friend who admired *Walden:* "I shall cause it to be known to him that you sat up till two o'clock reading his book; and he will pretend it is of no consequence, but will never forget it."

He made $96.60 in royalties, a nice haul, but it quickly went out of print. (Only a few weeks before he died, he finally convinced a publisher to reissue it, never to go out of print again.) It did not do as well as Thoreau had hoped it would. He tried to get his lecturing going again, Horace Greeley advertising him in the *New York Tribune;* he planned a trip to the West, sending out notices of his tour. He worked up a multi-lecture routine, beginning with "Moonlight," an essay based on the moonlight walks that were a fad among his Transcendentalist neighbors, and followed up by "Walking, or the Wild," "Getting a Living," and "What Shall It Profit." He landed a few dates; he delivered "Moonlight" in Philadelphia. It did not go over well. He returned via New York, stopping to see P. T. Barnum's museum and visiting his friend Greeley, who, Thoreau wrote, "appeared to know and be known by everybody." They went to the opera, Greeley attending for free. When Thoreau returned home, the western lecture tour was canceled, given that only two towns had responded: Akron, Ohio, and Hamilton, Ontario.

Even prepared for failure, he was struck by it. His emotional low points came seasonally—on the anniversary of the death of his brother—but now they were amplified, like a full moon tide. He wrote in the morning, walked, wrote in the afternoon, and ate at home most nights, with his family—his father, who was mostly deaf, sat at the head of the table, his mother and sister—and, for three years beginning in 1855, a Harvard student, Franklin Sanborn, and his sister, who, on Emerson's encouragement, had opened a local school. Sanborn eventually convinced Thoreau to take his students out on hikes in the woods. The kids loved him, and he loved the kids, who didn't think it odd that he was known as a man who couldn't pass a berry without picking it. They loved that he knew all the birdcalls. Still his friends chided him for his curiosity, to which the thirty-seven-year-old replied, "What else is there in life?"

His spirits affected his health, most likely, the tuberculosis causing him to fatigue more and more easily, the bacteria spreading through his lungs and his nervous system. (Once, folk medicine associated tuberculosis with vampires. When one person in the family died of it, it seemed as if the life had been taken from the rest of the family members, when in fact they were most likely all infected.) After *Walden,* he entered a protracted illness that made him unable to walk for a time; for weeks, he had trouble getting out of bed. He hated not walking around. For the first time in his life, his friends thought him lazy. He had planned to return to Maine, but called the trip off. He managed to explore in town, capturing a flying squirrel that he released to fly around in his room in the attic for a short time. Then, as his health slowly returned, his zest for natural details increased. He studied nests. He collected information on tree rings, on pollen; he began to create large charts, tables of information. He tapped maple sap for syrup. He made sap wines, using the sap of various birch trees, consulting with Alek Therien, the Concord woodcutter, who knew alcohol better than he did. He studied mice for a time, then muskrats. He killed ducks, skinned them, described their anatomy, observing things closely, waiting, hoping to see connections. Thoreau was more and more curious about the flora and fauna and what we would today call the ecology of Concord. The critic Lawrence Buell has noted that after the first drafts of *Walden,* Thoreau refined his perceptions of the natural world around him. Thoreau noted the variety of apple blossom scents and began especially to notice microclimates in the area, such as the Miles Swamp, a place devoid of wandering cows and woodchoppers. Thoreau described Saw Mill Brook as a mountain stream without mountains. If he was a radical environmentalist, it is because he radically re-noticed things.

His illness continued throughout the summer. "I have

been sick and good for nothing but to lie on my back for two or three months and wait for something to turn up," he wrote. His downturn was most likely related to the chronic tuberculosis—Channing noted that his friend's cough was particularly bad. The Emersons tried in vain to convince him to recuperate at their house. He accepted, but then turned down, an offer to live with Horace Greeley on the Greeleys' farm in Chappaqua, New York, where he was to tutor Greeley's children; according to Walter Harding, Thoreau realized, first, that he could never again be happy teaching, and, second, that Concord was his life and his work now. Finally, as summer ended he began to feel better. He took a trip to Cape Cod, taking notes, a working convalescence. "After four or five months of invalidity and worthlessness, I begin to feel some stirrings of life in me," he said. With the stirring he wrote the series of articles that became the book *Cape Cod*.

CAPE COD CAN SEEM LIKE A MERE TRAVEL BOOK, a lighthearted account of a trip along a stretch of ocean, but it is not so much travel literature as a dark tonic to *Walden*'s morning call. Whereas *Walden* sings of rebirth, life reimagined, *Cape Cod,* on the surface funny and wry, is a book in which the ocean waves roll in with death and destruction. It was published as magazine excerpts, in *Putnam's,* and it is not clear if even the editor knew it was to be a book. It was also about a region that seemed off the beaten track of New England history, the raised fist of the Cape's coastline itself a kind of parenthesis, as Thoreau would have noticed. It was considered wild and unkempt, not the vacation destination it is today. In other words, if you wanted to write a big book, a seller, you wouldn't want to write a book about Cape Cod. It was the opposite of a quick beach read.

Thoreau, though, had found another faraway place within a short train ride of Boston, a trip to the source of the Nile but as close as Buzzard's Bay. After his second trip to Cape Cod, during his long physical malaise, he contracted with G. W. Curtis, a friend and Brook Farm alumnus who had come from the farm to help Thoreau raise the walls of his house at Walden and was now working at *Putnam's.* Thoreau was always nervous about editors, who were often cutting passages behind his back, but he was, as always, interested in publishing; so in July, he and Ellery Channing made another trip to Cape Cod, to freshen up his reporting.* Sure enough, *Put-*

* James Russell Lowell, the *Atlantic Monthly* editor, had cut a mention of a pine tree and heaven—"It is as immortal as I am and perchance, may go to as high as heaven, there to tower above me still"—and Thoreau wrote to say he was "mean and cowardly." Thoreau is often criticized for not being pleased when his writing was rewritten, and responding indignantly, and his reaction is added to the pile of proof that he was unsocialized. I find it inspiring that he held to his standards and yet managed to be published anyway. He managed to make enough money to survive (and help support his family, even though it is often noted that he didn't have a family of his own and a mortgage to bear) while at the same time he kept hold of his writing, crafting it further and waiting for a receptive editor. His backup plan—never publishing—paid off, in that he was eventually published. He could afford not to publish because he did not depend on his writing for financial survival. He went the long way around. It was difficult. Thoreau scholar Steven Fink points to a letter Thoreau wrote in 1848, when he could find no one to publish *A Week,* to James Elliot Cabot, editor of the *Massachusetts Quarterly Review:*

> My book, fortunately, did not find a publisher ready to undertake it, and you can imagine the effect of delay on an author's estimate of his own work. . . . I esteem it a rare happiness to be able to write anything, but there (if I ever get there) my concern for it is apt to end. Time & Co. are, after all, the only quite honest and trustworthy publishers that we know. I can sympathize, perhaps, with the barberry bush, whose business it is solely to ripen its fruit (though that may not be to sweeten it) and to protect it with thorns, so that it holds on all winter, even. . . . But I see that I must get a few dollars together presently to manure my roots. Is your journal able to pay anything, provided it likes an article well enough?

nam's eventually pulled the plug on *Cape Cod,* afraid a section on an oysterman might offend for its salty, un-God-fearing language. But the first two sections were published, and like a dark joke, the book began in the manner of the typical genial travel essay: "I have been accustomed to make excursions to the ponds within ten miles of Concord, but latterly I have extended my excursions to the seashore."

Rather than living in a bucolically peaceful coexistence with nature, the residents of the Cape battle the churning, land-destroying, ever-recycling sea. Nature generally, in the form of salt and waves and sand, is oblivious of the man who battles to live on a barren spit of Cape Cod. To Thoreau, the arm of the Cape is like the arm of a boxer, and life on Cape Cod is hostile, even for the land itself. "As I looked over the water, I saw the isles rapidly wasting away, the sea nibbling voraciously at the continent, the springing arch of a hill suddenly interrupted, as at Point Alderton,—what a biologist might call premorse,—showing, by its curve against the sky, how much space it must have occupied where now was water only," he wrote. Not that death is itself an end. "On the other hand, these wrecks of isles were being fancifully arranged into new shores, as a Hog Island, inside of Hull, where everything seemed to be gently lapsing into futurity."

Stylistically, *Cape Cod* has more in common with the New Journalism of the 1960s than it does with antebellum travel narratives: excursions within excursions, an obsession with facts, oral history, and lists, not for their own sake but for the sake of the darker meaning. If *Walden* is a call to life; *Cape Cod,* despite being sold alongside postcards in gift shops, was written as a study of death, an analysis of life's contradictions. Nature could inspire morality, but Thoreau has, more terrifyingly, also discovered an amorality in nature. In the detritus that the sea "vomits" on shore, Thoreau could see decay, a fact that worried him but that he neither ignored nor prettied up. It's not as if his life changed in any one way to make such a dark book. Sure, he was a little down about *Walden* not flying off the shelves, but the new book was part of his idea of confronting things, of being sincere, and he wanted to look at the dark side of the world he had assigned himself to continually monitor. The audience would have expected him to whistle past the grave, to talk about pearly gates, but he was doing close-ups, as if opening some noir TV cop show.

Meanwhile, the book is funny, full of black humor, especially in the way it pokes fun at travel writing; it's easy to imagine the passages that would be well-received at a lecture—as he describes *not* seeing the town of Sandwich, for example: "Ours was but half a Sandwich at most, and that must have fallen on the buttered side some time." Or as he pokes fun at the teeth of the residents, and his own (he had very recently had all his teeth pulled):

> A strict regard for truth obliges us to say, that the few women whom we saw that day looked exceedingly pinched up. They had prominent chins and noses, having lost all their teeth, and a sharp W would represent their profile. They were not so well preserved as their husbands; or perchance they were well preserved as dried specimens. (Their husbands, however, were pickled.) But we respect them not the less for all that; our own dental system is far from perfect.

The central characters of the book are the wreckers, mining the shore for the flotsam and jetsam. Thoreau goes back and forth on who they are. They are painted as getting something from nearly nothing, beach-based self-made men, or as parasites that prey on everything and everyone. As the latter, they are capitalist parasites—another shot at America's obsession with business. The wrecker fascinated Thoreau and horrified him, as he could see that he was a wrecker himself, as all writers can be, writing of someone else's life, rather than his own, as he had, at least, in *Walden:* "But are we not all wreckers contriving that some treasure may be washed up on our beach, that we may secure it, and do we not infer the habits of these Nauset and Barnegat wreckers, from the common modes of getting a living?"

It's great to see it in the racks in gift shops at the beach, but as it happens, *Cape Cod* is Thoreau's book about America, conducted in the opposites and paradoxes that he loved. While everyone in America was racing west, he headed east. He backtracked. While the political parties were preaching Manifest Destiny, claiming the righteousness of American expansionism, Thoreau described the latest Americans washing up dead on a shore. The analysis behind *Cape Cod* is not something that would get him a lot of votes in a political campaign even today, and in his day it meant editors were reluctant to work with him; the *Cape Cod* editor eventually dropped him. In his rereading of history, America wasn't so much discovered as stumbled into by one semi-suspecting group after another, with and without the Puritan God. It's a place where the hopeful, the enslaved, the conniving, have come to make a new life or re-create an old one, sometimes successfully, sometimes not, sometimes right alongside each other. From the Cape, Thoreau sees the wrecks of the Pilgrim-discovered America alongside all the others, explorers united in death and the passage of time. The Vikings, the Spanish, the Basque, the Dutch, the French had all been to the shores of New England before Columbus, before Plymouth Rock.

In the twenty-first century, this is a little-less-startling way to read the American past. It was scandalous in Thoreau's time. It spits in the face of Manifest Destiny. Still, he saw the American past as proof that life could renew. As morbid as it is, the wreckage inspires a kind of hope, if not a challenge, a call to a new revolution or, in the call of the Transcendentalists, *real* reform. "If America was found and lost again once," Thoreau wrote, "as most of us believe, then why not twice?" *Walden* calls us to jump out of the boat as we race over the falls of greater and greater materialism. *Cape Cod* is the book that looks back at the source of it all, that shows us we are not coming from where we thought we were coming from. Here Thoreau ends up looking less like a dead white male and more like the original alternative historian. Especially when it comes to Indians. A passage that stands out as forward-thinking, even by today's standards, when it comes to Native American treaty rights, is this one, which is also funny, very Mark Twain (who was working as a printer when Thoreau wrote it):

> When the committee from Plymouth had purchased the territory of Eastham of the Indians, "it was demanded, who laid claim to Billingsgate?" which was understood to be all that part of the Cape north of what they had purchased. "The answer was, there was not any who owned it. 'Then,' said the committee, 'that land is ours.' The Indians answered, that it was." This was a remarkable assertion and admission. The Pilgrims appear to have regarded themselves as Not Any's representatives. Perhaps this was the first instance of that quiet way of "speaking for" a place not yet occupied, or

at least not improved as much as it may be, which their descendants have practised, and are still practising so extensively. Not Any seems to have been the sole proprietor of all America before the Yankees. But history says, that when the Pilgrims had held the lands of Billingsgate many years, at length, "appeared an Indian, who styled himself Lieutenant Anthony," who laid claim to them, and of him they bought them. Who knows but a Lieutenant Anthony may be knocking at the door of the White House some day? At any rate, I know that if you hold a thing unjustly, there will surely be the devil to pay at last.

Cape Cod ALSO SHOWS US how life had changed in Concord for Thoreau, and for his neighbors, after *Walden*—in fact, how life had changed in metropolitan areas throughout America: a flood of immigrants had arrived. Irish immigrants open *Cape Cod,* the victims of a shipwreck, their bodies strewn across the shore. It's a harrowing beginning to a seemingly leisurely travel piece, a smack in the face. He writes about what is largely not written about in American literature at the time: the invisible poor. At the time, the Irish were arriving in America in larger and larger numbers. They were fleeing An Gorta Mór, "the great hunger," or as it is more commonly known, the Irish Famine, and, in the thinking of Thoreau's contemporaries, infecting America.*

* Despite Thoreau's writing and his status as a founding father of environmentalism, a considerable amount of xenophobia is sometimes generated in environmental circles. Betsy Hartmann, the director of the Population and Development Program at Hampshire College in Amherst, Massachusetts, has written about what she has described as "the greening of hate." "The greening of hate—blaming environmental degradation on poor populations of color—is once again on the rise, both in the U.S. and overseas," she writes. "In the U.S., its illogic runs like this: immigrants are the main cause of overpopulation, and overpopulation

The years of the emigration roughly coincided with the years that Thoreau began and finished *Walden;* between 1846 and 1855, about 1.6 million Irish refugees arrived in the United States, which is near the total number of all the immigrants who had come during the previous seventy years. More than a million people died in the famine, and a million left Ireland, while the British government debated whether to provide charity or allow market forces to feed and house the poor. Vast numbers settled in Boston; by the time Thoreau published *Cape Cod,* nearly one third of all Bostonians were Irish immigrants. (In New York, in the 1850s, 30,000 Irish men, women, and children were reportedly living in cellars, without light or drainage.) Aside from laying the railroad that ran through Concord, the Irish found other work—digging ditches, running the mules that turned the spinning wheels in the textile mills, cleaning the homes of Boston's middle-class families, working in boardinghouses, as they did at the Thoreaus'; they worked jobs where, like Hispanic immigrants today, they were routinely exploited. As a labor pool, they helped move the East Coast cities from commercial to industrial economies, the economic transition that Thoreau critiqued with *Walden.* As a cultural group, they were abhorred, considered, in the words of Boston mayor

---

in turn causes urban sprawl, the destruction of wilderness, pollution, and so forth. Internationally, it draws on narratives that blame expanding populations of peasants and herders for encroaching on pristine nature. In the first instance, the main policy 'solution' is immigration restriction; in the second it is coercive conservation, the violent exclusion of local communities from nature preserves. Both varieties of the greening of hate are about policing borders. By stressing the negative role of population growth, both target poor women's fertility as the fundamental root of environmental evil." In the United States in the mid–1990s, anti-immigrant groups attempted to reduce immigration in the name of environmental protection.

Theodore Lyman, "a race that will never be infused into our own, but on the contrary will remain distinct and hostile." *Crania Americana,* a racist ethnography, said, "Their wild look and manner, mud cabins and funereal howlings recall . . . a barbarous age."

Thoreau has a mixed record as far as his acceptance of the immigrants as human beings. In *Walden,* he painted the Irishmen as lowlifes, the foil to his call for a higher living. His puns were in line with cultural stereotypes; as they grade the railroad, the Irish were continually "degraded." Thoreau describes John Fields, a worker living in a shanty at Walden Pond with his wife and children, as "shiftless." "The culture of an Irishman is an enterprise to be undertaken with a sort of mortal bog hoe," Thoreau wrote in *Walden.* There is evidence that as drafts of *Walden* proceeded, as his own exposure to the immigrants increased, Thoreau may have tried to soften his portrait, but not by much. His journal is a different story. In the journal, the Irish changed after 1850. He admired their fence-making skills: the sod fence was cheaper and more ornamental than a stone fence. He admired their conscientiousness, the small ways they scraped together the sums required to bring their families over. He called a guy named Flannery the hardest-working man in town. He described helping clothe an Irish boy whom he admired for his perseverance in getting to school, in the snow, with little to wear. (The boy's older sister, thirteen-year-old Catharine Riordan, worked in the Thoreau kitchen.) Slowly, and to his neighbors' chagrin, Thoreau became an advocate for the Irish in Concord. He argued against legal attempts in Concord to prevent the Irish from searching for firewood. He wasn't always their biggest defender; he made fun of Emerson's gardener, an Irishman named Hugh. But he changed his mind, or opened it, an example of him practicing what *Walden* preached. Once, after an Irish worker's employer extorted from him a four-dollar cash prize he had won at a fair, an incensed Thoreau went around Concord to raise the money for the man, paying off the employer. Long after he seemed to have given up on marriage, he would have a crush on an Irishwoman working at a friend's home: she had read his work and he saw a true lover of nature, or at least of his writing, a common occupational temptation. He even adopted the habit of wearing corduroy from the Irish workers, to his non-Irish neighbors' disgust.

It should not be a surprise that Thoreau had thought about the Irish Famine. It is likely that Margaret Fuller's ideas about the Irish might have rubbed off on him too. Her writing on the Irish immigration for Horace Greeley's *New York Tribune* can sound patronizing today, but it stood out at the time. "Their virtues are their own; they are many, genuine, and deeply rooted," she wrote in a column in 1845, entitled "The Irish Character." "Can an impartial observer fail to admire their truth to domestic ties, their power of generous bounty, and more generous gratitude, their indefatigable good humor, . . . their ready wit, their elasticity of nature." She went on to say, "They are fundamentally one of the best nations in the world. Would they were welcomed here, not to work merely, but to intelligent sympathy, and efforts both patient and ardent, for the education of their children!"*

* Often when Fuller spoke of Irish people, she spoke of Irishwomen, in particular those working as nannies, and asked her readers to forgive them their shortcomings, given the work they did. In one column, Fuller described a nanny sweating it out on a hot day with a baby, while the family that belonged to the baby was on a horse carriage or a buggy ride of some kind. "They [the family] were having a pleasant time; but in it she had no part, except to hold a hot, heavy baby, and receive frequent admonitions to keep it comfortable. No in-

---

AT WALDEN, OR AT THE WALDEN THOREAU would have you imagine for yourself, life happens—a chrysalis hatches within the deadwood. At Cape Cod, as in the world of pre–Civil War industry, death happens, and the wreck of the ship in the opening of *Cape Cod* brings home America's dark side. "DEATH! ONE HUNDRED AND FORTY-FIVE LIVES LOST AT COHASSET," reads the handbill that Thoreau picks up in Boston, waiting for the ferry to Cape Cod, which is late due to deadly weather. If it had been travel writing that he was doing, he might have avoided writing about dead people altogether. But *Cape Cod* is looking at the engine of American progress, and here is some engine trouble. When he got the handbill, Thoreau was with his friend Channing, and on the train to the Cape they were accompanied by relatives of the dead. As they set out on the beach, they came to the scene of the wreck of the *St. John,* a boat that had left Ireland on September 7, 1849. On November 7, a storm had driven the *St. John* into the cape. It was just one of about sixty "coffin ships," as they were known, that had been lost between 1847 and '53. He reports:

> The brig *St. John,* from Galway, Ireland, laden with emigrants, was wrecked on Sunday morning; it was now Tuesday morning, and the sea was still breaking violently on the rocks. There were eighteen or twenty of the same large boxes that I have mentioned, lying on a green hill-side, a few

---

quiry was made as to her comfort; no entertaining remark, no information of interest as to the places we passed, was addressed to her." Fuller added, "Her joys, her sorrows, her few thoughts, her almost buried capacities, would have been as unknown to them, and they as little likely to benefit her, as the Emperor of China."

> rods from the water, and surrounded by a crowd. The bodies which had been recovered, twenty-seven or eight in all, had been collected there. Some were rapidly nailing down the lids, others were carting the boxes away, and others were lifting the lids, which were yet loose, and peeping under the cloths, for each body, with such rags as still adhered to it, was covered loosely with a white sheet. I witnessed no signs of grief, but there was a sober despatch of business which was affecting. One man was seeking to identify a particular body, and one undertaker or carpenter was calling to another to know in what box a certain child was put.

For a long time, this passage has been used to show how heartless Thoreau was. I think that's wrong. Again, I'm not saying he couldn't be a jerk; Emerson was thinking of him as more and more of a jerk by this time. He was ultimately a kind of Robin Hood of jerkiness. A jerk, for certain, to pretentious intellectuals, and less so, sometimes, to farmers, whom he probably wanted to emulate. This passage is significant in the annals of American immigration history in that it exists. In the American culture at large, few cared about the Irish, but here Thoreau bears witness and, in so doing, shows a compassion that most Americans did not have for the latest group of Americans-to-be.

Thoreau's vantage point is as a naturalist-philosopher, with a view of nature that has decay at the roots of life. He also sees the event in the context of a repertoire of past atrocities he has witnessed, such as the death at sea of Margaret Fuller, for instance, just a few autumns before. As mentioned, in 1850, a few weeks after his second trip to Cape Cod, Thoreau went to New York in hopes of retrieving the body of Margaret Fuller from amidst the wreckage of the *Elizabeth,* the ship carrying Fuller, her husband, and their child from Italy, where Fuller had been writing about the Italian revolution for the *New York Tribune.* The *Elizabeth* had struck a reef just off Fire Island; its captain had died of smallpox, and the first mate, who had taken over, misjudged the ship's location after a hurricane. Emerson advanced Thoreau money and asked him also to look for Fuller's personal belongings, including a manuscript. By the time Thoreau arrived on the beach at Fire Island, a few days later, the scene was similar to the beach at Cape Cod. Wreckers combed through the debris. People came to identify bodies. Graves were being dug. Thoreau saw skeletal remains but did not find Margaret Fuller, though he heard of her last minutes from the first mate—they could not get the lifeboat off. Only her child, Angelino, washed ashore dead.

Thoreau had seen his share of violent death. In January 1853, at the anniversary of his brother's death and while in the midst of a deep funk, Thoreau had also been a witness to the aftereffects of a violent explosion at a gunpowder factory, near his home in Concord. In his journal, he described the effects to the buildings and then at last to the workers, whose bodies were strewn across the nearby fields: "The bodies were naked & black—Some limbs & bowels here & and there & a head at a distance from its trunk. The feet were bare—the hair singed to a crisp." The scene haunted him, and he mentioned it repeatedly in his journal, a few weeks later describing a nightmare in which he dreamed he unearthed and touched dead bodies, feeling defiled afterward. Early the following summer, he noticed bits and pieces of the destroyed mill floating down the river, and wrote, "How slowly the ruins are being dispersed!" Imagery reminiscent of the description used in his journal entries on the explosion found its way into the final drafts of *Walden,* in the sand-cut passage, where images of decay and dismemberment manage to give way to life. In his journal, he obsessed over the systematic differences between a body that is alive and one that is not, between himself alive, himself dead. Returning from the scene of Margaret Fuller's death, he wrote, "Who knows but you are dead already?"

On Cape Cod, he looks closely at the dead immigrants, women mostly:

> I saw many marble feet and matted heads as the cloths were raised, and one livid, swollen, and mangled body of a drowned girl,—who probably had intended to go out to service in some American family,—to which some rags still adhered, with a string, half concealed by the flesh, about its swollen neck; the coiled-up wreck of a human hulk, gashed by the rocks or fishes, so that the bone and muscle were exposed, but quite bloodless,—merely red and white,—with wide-open and staring eyes, yet lustreless, dead-lights; or like the cabin windows of a stranded vessel, filled with sand. Sometimes there were two or more children, or a parent and child, in the same box, and on the lid would perhaps be written with red chalk, "Bridget such-a-one, and sister's child." The surrounding sward was covered with bits of sails and clothing. I have since heard, from one who lives by this beach, that a woman who had come over before, but had left her infant behind for her sister to bring, came and looked into these boxes, and saw in one,—probably the same whose superscription I have quoted,—her child in her sister's arms, as if the sister had meant to be found thus; and within three days after, the mother died from the effect of that sight.

It's like a crime scene photo, cold and cruel, but not for nothing, and Thoreau sees that the dead Irishwomen, who had come to be servants, have made the transition that he pondered on returning from Margaret Fuller's drowning, when he wondered whether we were all alive or dead and what the difference is. "I saw their empty hulks that came to land; but they themselves, meanwhile, were cast upon some shore yet further west, toward which we are all tending, and which we shall reach at last, it may be through storm and darkness, as they did," he wrote.

Today the federal government builds a wall along the border with Mexico. In Thoreau's day, the first immigration stations were set up to inspect Irish immigrants, in order to repel what the governor of Massachusetts compared to the "horde of foreign barbarians" that brought down the Roman Empire. The Irish were thought to carry disease, as well as the greater risk of miscegenation. Prominent citizens such as Harriet Beecher Stowe's father, Lyman Beecher, and Samuel F. B. Morse railed against the Irish taking over the country from the "native Americans." In American literature, the Irish Famine goes nearly unmentioned. (Herman Melville's novel *Redburn* describes the horror of a famine ship in transit, and Melville saw that immigration was what made the United States a new nation, rather than a carbon copy of Europe—"not a nation, so much as a world.") That's partly because of the racism and anti-Irish xenophobia of the time. It's partly because the famine and the cargo ships full of dying immigrants stands in relief to all that Manifest Destiny stood for: the national prosperousness. Put another way, the feminine issues associated with famine—hunger and family—are at odds with the masculine dreams of expansion and conquering. Again, Thoreau describes what is nowhere else described.

And death is the order, or the disorder, the frightening anarchy that stands in contrast to the idea of progress in America. The girl without a name—Bridget such-a-one—stands for diaspora. Her namelessness is not a slight; it's a device.* They are not savages; they are us. The description of the nudity borders on fetishistic, but nobody said Thoreau wasn't odd. But in this case, near nakedness shows us just how bad things have gotten: it's a political breakdown, the center not holding, an opening of a horror film. Thoreau sides with the dead: "The strongest wind cannot stagger a Spirit; it is a Spirit's breath. A just man's purpose cannot be split on any Grampus or material rock, but itself will split rocks till it succeeds." And, most democratically, Thoreau sees these border crossers as New Englanders, as Americans, as the most recent arrivals in the long line of explorers—as young women who are, as he writes, "coming to the New World as Columbus and the Pilgrims did."

Thoreau was no saint. In significant ways he maintained his biases—the pro-Thoreau part of me wants to think that he was smart enough not to treat the Irishmen like idiots when he first met them, but that is not the case. But he took the Irish into America's story at a time when the land of op-

* That Thoreau refers to the name on a coffin as "Bridget such-a-one" in these passages of *Cape Cod*, not mentioning her surname, rankles Irish American historians. Irishwomen working as housecleaners and nannies were often referred to generically as "Bridgets" or "Kathleens." But it's significant that he names her at all. It's also significant that he identifies a woman. The Irish were unusual among immigrant groups in that the women outnumbered men; they were the young on their way, as he notes, to work as maids and nannies—to take jobs that, as Hasia Diner writes in *Erin's Daughters in America,* most other women turned down.

portunity began to work hard to shut people out—in this case, the starving poor. He stops at the edge of the world, with the nation at his back, and sees that America is like its coast, battered by the waves, life in a battle to maintain itself, death consuming all, eventually granting citizenship to everybody.

IN MANY WRITERS' LIVES, there is the work that the writer wants to do and the work that the writer must do, to stay afloat. After *Walden,* Thoreau managed to combine the two in a lot of ways through what became his primary line of work, surveying. He still wrote with a ferocious diligence, especially in his journal, and occasionally wrote essays, but he wrote at times as if it were an avocation. His larger projects—on leaves and wild fruits, on trees and tree rings, and on Concord itself (a calendar, something along the lines of his beloved *Georgics* that would document the day-to-day natural life of Concord, all its cycles, human and otherwise)—were writing projects that required the observation that a surveying job could support.

At the very least, surveying helped him get around, in Concord and the East Coast generally. In 1856, he surveyed in Eagleswood, New Jersey, an Associationist community; they were hoping to make it into a place for commuters, with good schools and lectures, within steamboat distance of New York. He ended up laying out vineyards and surveying for individual owners, attending dances with the children, even though he didn't feel like dancing on the night he was invited. "They take it for granted you want society!" he wrote to his sister. Thoreau liked to dance only when he liked to dance, as many people can understand. He also lectured them, at their urging. "It was expected that the spirit would move me (I having been previously spoken to about it); and it, or something else, did,—an inch or so. I said just enough to set them a little by the ears and make it lively." They loved him, even though he read the hard-on-the-audience lecture "What Shall It Profit" (eventually renamed "Life Without Principle"). "I think that there is nothing, not even crime, more opposed to poetry, to philosophy, ay, to life itself, than this incessant business," he said. Afterward, he wrote to his sister to say that he missed Concord.

In 1857, the U.S. economy again soured. During the summer, banks began to fail; the ones that remained solvent suspended currency payments. The panic turned into what would later be known as the Depression of 1857. In November, Thoreau wrote his friend H. G. O. Blake:

> The merchants and company have long laughed at transcendentalism, higher laws, etc., crying "None of your moonshine," as if they were anchored to something not only definite, but sure and permanent. If there was any institution which was presumed to rest on a solid and secure basis, and more than any other represented this boasted common sense, prudence and practical talent, it was the bank; and now those very banks are found to be mere reeds shaken by the wind. Scarcely one in the land has kept its promise. . . . Not merely the Brook Farm and Fourierite communities, but now the community generally has failed. But there is the moonshine still, serene, beneficent, and unchanged. Hard times, I say, have this value, among others, that they show us what such promises are worth,—where the *sure* banks are.

WHEN HE WAS DONE SURVEYING, or after he spent a few days on a money-making project, he always came back to the journal. He had turned his journal into something else, something more than even Emerson understood until after Thoreau was gone and his friends began passing it around—each amazed as they read, each, in a way, flabbergasted, for the journal is huge: two million words. It is the result of almost daily work. And yet even today, it stands in *Walden*'s shadow.

The voice of the journal is different from the voice of the narrator of *Walden,* in intent and in the popular mind. The journal is Thoreau himself, and when you look at the journal, you see not only the author, a version of the man we think we meet in the artificially constructed *Walden,* but the stuff around him too—the place he walks, works, and lives. For his entire life he honed his skills as an observer, relishing facts. *This* nature—the nature in which he lived his every day—is not what we would today think of as a wilderness or an Elysian Fields. It was a place where people live or get their living.

Surveying was his fieldwork. At various times, he had attempted to become a house painter, a speculator (in berries), and a fence maker, but he excelled in surveying, and was prized for that skill. He was considered scrupulous; his work in settling boundary disputes often helped two parties avoid court, and he worked with courts in settling land claims in estate cases. In a dispute between Emerson and a landowner named Charles Bartlett, Bartlett, who knew Thoreau was friends with Emerson, trusted Thoreau to work everything out. Thoreau used his surveying skills to set up the field for the Plowing Contest at the Middlesex County Fair. Each year, as the official surveyor for Concord, he would travel the circumference of the village.

Despite his reputation as a hermit, he was enmeshed in

the real estate business as a surveyor: he designed roads for local factories and helped the railroad lay out a new street to the depot. His largest survey project was called the River Project and executed on behalf of the Sudbury and East Sudbury Meadow Corporation in their controversy with the Middlesex Canal owners in 1860. He collected information relating to the water depths in the Concord River at each bridge from Wayland to Billerica, and drew up charts and drawings after consulting with officials and landowners in several towns, which were later used in court. James B. Wood, a Concord lumberman, paid Thoreau $3 a day to survey woodland and described him as "always very pleasant, talkative, and ingenious."

But tuberculosis was creeping up on him, the bug spreading. It became more painful to breathe, and he tired more frequently, though he still managed his work and walks. He grew hair on his neck and face, to fight off the colds that exacerbated his tuberculosis, and if he didn't look homely, then he began to look moss-covered, a stump older than its age. Night sweats, loss of energy and appetite, were followed by bouts of what seemed like good health. Despite the increasing amount of time it took up, surveying was still his emotional part-time job, albeit the paying part. His full-time job was observing and describing and often experiencing the plants and animals and human doings around him, surveying life: once he found a bottle of blackstrap rum left behind by mowers and, on tasting it, found it still had kick. Because he was convinced he could learn everything he needed to know about himself by seeing the world around him, he set out to exhaustively document that world—*everything* in Concord. This did not mean that he was off alone in the woods, though often he was. The woods, or what was left of them at the time, as well as the fields that were predominant, were populated, and to document their activities meant interacting with loggers, farmers, and fieldworkers who worked there.

"It is in vain to dream of a wildness distant from ourselves," he wrote. "There is none such. It is the bog in our brain and bowels, the primitive vigor of Nature in us, that inspires that dream. I shall never find in the wilds of Labrador any greater wildness than in some recess in Concord, *i.e.* than I import into it."

WHAT WAS THOREAU'S NATURE? What was the place he traipsed around in every day like? The landscape that Thoreau was describing more and more minutely as his life came into its final years had itself changed drastically during the 250 years prior to Thoreau's explorations. Settlers had spread across it in colonial times, in the wake of an Indian population destroyed by disease and war. The colonists began to set up farms, so that by the time Thoreau's family moved back to Concord, sixty percent of New England was open fields; only woodlots remained. In Thoreau's father's time, farmers and craftspeople lived off the land in self-sufficient communities, but by the time Thoreau was an adult a new phase was about to begin: the 1830s, when, simultaneously, trains arrived, people began moving west, farms were being abandoned, and fields were reverting back to woods. (By 1900, a new logging industry would develop in New England, logging the pine woods that grew in the old fields.) When Thoreau was at Walden Pond, the New England landscape was near its peak of deforestation. It was being farmed, mined, and logged more intensively than ever. It was by no means a pristine landscape or what we might think of today as a "natural" landscape. It was, to use the phrase of the ecologist David R. Foster, "a cultural landscape," a landscape that, like almost all landscapes that the average person sees, is shaped by the interaction of the natural environment *and* human history. The land had us all over it, in other words.

As I was trying to show in my exegesis on the sandbank passage in *Walden,* Thoreau was inspired as much by man's interaction with nature as he was by what we might call pure nature. Thoreau savored the view of not just the untouched top of Mount Katahdin (which would not exactly have been untouched, but let's say it was) but the vistas colored by crop selection, the sounds at dusk as they were punctuated by the echoes of chopping wood. He was excited by the human activity that surrounded him. Even the brilliant landscape historian J. B. Jackson has his Thoreaus confused, as far as I can tell. "Far from discovering virtues in the agrarian way of life, Thoreau was scarcely more tolerant of farmers than he was of city dwellers," Jackson wrote. As proof, Jackson cites this line from *Walden:* "The farmer leads the meanest of lives. He knows Nature but as a robber."

But here Jackson gets Thoreau's tolerance wrong on both counts. It is farming *practices* that Thoreau protests in his journal, not the farmer, not the farm. In *Walden,* of course, the narrator is a pose, and this is the Yankee huckster, having fun with the farmer, mostly in hopes of changing the relationship between farmer and farm, the economic forces that pushed a farmer to do something other than farm for himself and his community—it's the same story that the family farm faces today, as it struggles alongside commercial farming. For Thoreau, the opposite of living "meanly" is living "sincerely," in a relationship that is mutually beneficial, a relationship that is just and practical, rather than damned from the get-go. "The farmer increases the extent of the habitable earth," wrote Thoreau. "He makes soil. That is an honorable occupation." Indeed, the farmer was Thoreau's hero.

In the journal, we see a nature that is more complicated than a landscape calendar of *Walden* photos, and perhaps just as interesting, if not more varied. Thoreau watched the trees disappear. "Every large tree which I knew and admired is being gradually culled out and carried to a mill," he wrote. There were fewer large animal species—the moose had been nearly exterminated—but there were animals, and each year he detailed the cattle drives from one part of town to the next. He saw the sleighs used in the winter abandoned alongside the road in the summer. Always, the reader of Thoreau's journal hears things—rarely is it quiet at Walden or anywhere in Concord. Aside from the music of birds, there is the music of fieldworkers singing, of a piano being played in a parlor. In the winter, the woods echoed with thumps as boys clubbed the trees for chestnuts. The sawmills in the area, powered by water, ran day or night, whenever high water allowed. "Hear the sound of Barrett's sawmill, at first like a drum, then like a train of cars," he writes. In the calm summer, he noticed recreation. "Now I see gentlemen and ladies sitting at anchor in boats on the lakes in the calm afternoons, under parasols, making use of nature." In his view from a hilltop was a patchwork of fields bounded by fences, denuded slopes, grasses along streams that were hayed for winter food for livestock. Thoreau found it all thrilling. Then again, Thoreau found nearly everything in his IMAX view thrilling, down to the color of the fields, a sensibility to be envious of, it seems to me:

> These are the colors of the earth now: all land that has been sometime cleared, except it is subject to the plow, is russet, the color of withered herbage and the ground finely commixed, a lighter straw-color where are rank grasses next water; sprout-lands, the pale leather color of dry oak leaves; pine woods, green; deciduous woods (bare twigs and stems and withered leaves commingled), a brownish or reddish gray; maple swamps, smoke color; land just cleared, dark brown and earthy; plowed land, dark brown or blackish; ice and water; slate-color or blue; andromeda swamps, dull red and dark gray; rocks, gray.

THE LONGER HE LIVED, THE MORE INTENSELY—the more joyfully—he watched his neighbors as they interacted with his town; it's as if the life sucked from him by tuberculosis was somehow transferred to his humor, to his outward-reaching soul. Conservation didn't exist as a practice; it was more a matter of common sense, though it was not entirely common. Coming into the last years of his life, he began to see that nature was managed by people, to good and bad effect, and the balance sheet was affected in ways that were difficult to notice. He saw a dearth of primitive wood—what today we would call old-growth forests—recognizing them as places with a diversity of plants, as ecologically valuable. Even then, before industrialization and suburban sprawl, he characterized few forests in Massachusetts as being older than a few hundred years. His understanding of the woods is prescient: he suggested that a woodland area, in light of the rapaciousness of loggers, be preserved, for the sake of the community, while at the same time he extolled particular logging practices.

He admired his neighbor, George Minott, for his forest

management skills. "Minott tells me that his and his sister's wood-lot together contains about ten acres and has, with a very slim exception at one time, supplied all their fuel for thirty years, and he thinks would constantly do so . . . He knows his wood lot and what grows in it as well as an ordinary farmer does his corn field." The person who collected his own wood enjoyed his wood in a way that Thoreau romanticized: in scouting it, finding it, cutting it, or finding scrap wood or chips—in lugging it, each chip and log having a story, or so he felt. "Some farmers load their wood with gunpowder," he wrote, "to punish thieves." Emerson, he noted, burned twenty-five cords of wood, while Goodwin, a fellow woods-walker, burned a cord and a half, "much of which he picks out of the river." By the time Thoreau died, he could look over a field or a forest and estimate its history—the impact of old windstorms, the number of cuttings, whether it had been a field or a woodlot or, on rare occasions, a forest forever.

"Thank god they cannot cut down the clouds," he once wrote.

THOREAU READ CHARLES DARWIN EAGERLY, a few months after *The Origin of Species* was published, in 1859, and Darwin confirmed Thoreau's view of nature as a process. Thoreau began to abide by what he called a "developmental theory," in contrast to the creationist theory of Louis Agassiz, a naturalist, biologist, and geologist then at Harvard, for whom Thoreau had collected species, but less and less enthusiastically, skeptical as Thoreau was of his devotion to creationism. For Thoreau, the divinity of nature was in the creation processes—the unity behind Darwin's finches and the dispersal of seeds as he watched them grow in Concord's woods. Some critics have found it odd that Thoreau focused more and more on water temperatures, on counting tree rings and leaves, on the minutiae of his neighborhood, on the minutest of minutiae. But he was doing his version of Darwin. He had begun doing his Darwin, in fact, before he'd heard of Darwin, and he revved it up after, feeling confirmed, vindicated, as he plowed through the woods between surveying jobs, only his lungs slowing him down. On the first day of February 1859, he went to the Assabet River. He noticed that the river had suddenly gone down. On February 2, he watched a neighbor cut down an oak tree. On February 3, his father died. "He was quite conscious to the last," Thoreau wrote, "and his death was so easy that we should not have been aware that he was dying, though we were sitting around his bed, if we had not watched very closely."

As Thoreau grew older, he admired smaller and smaller processes, and he put his poetic powers to examining the links in things. Aside from charts and tables, he saw the snow braiding in the wind on the river, the ice crystals thawing in the dirt, flocks of sunlit particles dancing in the air. It was the combination of natural and human energy that propelled change in the landscape, and, looking for this, he saw the changes in the everyday. Apple trees were disappearing. Birds were not returning. Farms were getting bigger, smaller ones bought up, sometimes by city dwellers looking for rustic retreats. "We walk in a deserted country," he wrote as early as 1851. New Englanders were moving to mill towns and to the West, and farm fields were everywhere soon covered with shrubs, on their way to woods. He began to understand that the open meadow, a common sight of his youth, was the result of human activity. When he had graduated from college,

Concord was mostly open meadows. When his father died in 1859, a little more than two years before Thoreau would, the meadows were disappearing.

His understanding of the meadow was a crucial insight into the way the land works. In the meadow was a whole world. More and more in New England, but in other parts of the United States as well, the open meadow is mostly an image that we recall, an image we see in old landscape paintings, especially in the northeast; but as they have disappeared, our understanding of them has increased, or at least caught up to Thoreau's understanding: meadows and natural grasslands, or even seminatural grasslands—backyards, in other words—are understood by ecologists as critical if you are going to support a diversity of species. Thoreau's observations anticipate this understanding. In his journal, he noted the cycle of river meadows, as they were flooded, as grasses grew, as grasses were mowed by farmers, who flocked to them in the summer months like the bobolinks, meadowlarks, snipe, and plover in the months before the cutting. Farmers staked off their hay, sticks topped with bits of newspaper.

On the edges of the meadows and along fields, Thoreau saw the long fences as ecological storage points, as storage sites for seeds and saplings and species that rested while a field was used, most likely to return again, via the wind, or via foxes and squirrels. "The Great Meadows present a very busy scene now," he wrote in the summer of 1853. "There are at least thirty men in sight getting the hay, revealed by their white shirts in the distance, the farthest mere specks, and here and there great loads of hay, almost concealing the two dor-bugs that draw them and horse racks [sic] pacing regularly back and forth. It is refreshing to behold and scent even this wreck of meadow-plants." There was nothing pristine or separate about these scenes: the lethargy of field hands allowed one species of bush or tree an advantage while flooding prevented others, and slowly, as the economy changed and humans moved out of New England, the land changed again.

Chapter 11

AUTUMN

In 1860, at the end of his life, the farmers knew Thoreau and knew what he was talking about, and they were not at all averse to listening. He may have made his enemies in town—with the establishment types, mostly; certainly not with children, farmers, and older women who liked poetry. See him in the fall of that year, as he walked across town to the fair, where he was scheduled to deliver an address. See him in the beard that he barely trimmed. He was weaker at this point, thinner, tuberculosis about to finally take him, but he still walked as if healthy. "He wore straw hat, stout shoes, strong gray trousers, to brave shrub-oaks and smilax, and to climb a tree for a hawk's or a squirrel's nest," Emerson would say at his funeral, before two more harvest fairs would pass. Thoreau could be sullen and sad come the middle of every winter, but he was no sulker. He was happy to be at the fair. "His senses were acute, his frame well-knit and hardy, his hands strong and skillful in the use of tools," Emerson went on. "And there was a wonderful fitness of body and mind." See Thoreau shaking hands, meeting and greeting, tipping his tall hat, stopping to lean his tuberculosis-plagued body on his umbrella, passing through the assembled fair-goers, and noticing, always noticing—taking in the sundry details a solitary hermit would miss. "All sorts of men come to the cattle show," he wrote in his journal afterward. "I see one with a blue hat."

In his last years, Thoreau had mellowed in a lot of ways; he could still be stiff, of course, and he was always reserved. He could be quick to parry if you were a writer or a philosopher or a banker or Emerson, who complained still about Thoreau's military mind, always ready to battle. Often though, he writes warmly of the farmers and townspeople generally. He jots down the lingo of the boys in the street: "You don't know much more than a piece of putty." He adds it all into the journal, which as time goes on becomes the long Latin ode to everyday that he envisions, filled with the very *stuff* that makes up his home. The swamp waters are warm, and the sawdust from the mills, his cold toes have told him, has settled into the river bottom. As he walked into the fair, he had just planted four hundred pine trees on Emerson's lot at Walden, receiving $7.50 for two days' work. He was still not above manual labor (and it was probably not beneath him to hit Emerson with some wisecrack on his way out the door). Due to the death of his father, Thoreau was now in charge of the family graphite business, and as the CEO, he had changed to a finer method of grinding the graphite that he had invented, another profit improver. A few days before the fair, a big storm had hit Concord, the night so dark that when he went out for a walk in the streets, he whistled while he strolled, so as not to run into anyone. This was his last somewhat healthy year.

The address he read to the farmers that day at the fair was entitled "The Succession of Forest Trees," and it was more scientific than Transcendental, sprinkled with a charming and self-deprecating humor that he uses to pitch himself as the hobo of farms and fields. He is speaking to his heroes, after all:

> Every man is entitled to come to Cattle-Show, even a Transcendentalist; and for my part I am more interested in the men than in the cattle. I wish to see once more those old familiar faces, whose names I do not know, which for me represent the Middlesex country, and come as near being indigenous to the soil as a white man can; the men who are not above their business, whose coats are not too black, whose shoes do not shine very much, who never wear gloves to conceal their hands.

He equates himself to a crooked walking stick, an oddity, and suggests that it may look as if the fair's organizers have made a mistake by inviting him to speak. According to him, his qualifications are his interest, along with his appreciation of the crowd, the members of which he knows and admires not as a recluse but as a surveying socializer, who, he probably understands, is about to go away:

> In my capacity of surveyor, I have often talked with some of you, my employers, at your dinner-tables, after having gone round and round and behind your farming, and

> ascertained exactly what its limits were. Moreover, taking a surveyor's and a naturalist's liberty, I have been in the habit of going across your lots much oftener than is usual, as many of you, perhaps to your sorrow, are aware. Yet many of you, to my relief, have seemed not to be aware of it; and, when I came across you in some out-of-the-way nook of your farms, have inquired, with an air of surprise, if I were not lost, since you had never seen me in that part of the town or county before; when, if the truth were known, and it had not been for betraying my secret, I might with more propriety have inquired if you were not lost, since I had never seen you there before. I have several times shown the proprietor the shortest way out of his wood-lot.

Then, in reporting on a "purely scientific discovery," he sets out to explain to the audience something that he knows has puzzled farmers and puzzled him: why is it that pine trees are first to sprout in the newly abandoned fields and why do oaks generally follow? Guys like Emerson don't want to hear this. The farmers and field hands do. He went back to the notes he had taken on the farm fields as they were abandoned. He had noted the seeds carried by birds and squirrels and foxes. He had watched pastures grow scrub, and seen land formerly tilled convert slowly to pasture. Pine was less palatable to grazing animals, which continued to eat in a lot that a farm might have stopped farming. He noted too that squirrels and other animals used the pines as cover, and buried acorns. When, after twenty to forty years, the pine forest grew to a height worth logging, then oak trees replaced them. Oak trees succeeded pines. Thoreau was always willing to look at a long period of time in his ecological studies, not just one season but many, a willingness that distinguished him.

In this speech, Thoreau coined the term still used by foresters and ecologists today: *succession*. First came the pine, succeeded by the oak. Modern ecologists would later come to understand that the widespread existence of white pines covering New England at the start of the twentieth century was the result of chance, a coincidence having to do with the nature of plants and the nature of humans, of plant genetics and particular New England farming practices. In the 1920s, they would attempt to plant pine again, but it was in vain—the hardwoods kept coming back. To plant pine was to fight succession, though it would take them a while to discover what Thoreau had reported at the fair that day.

Thoreau was not a conservationist, *per se;* the job title did not yet exist. But he was on his way to becoming one. He recognized that Concord needed land-management practices similar to those followed in England. "Why not control our own woods and destiny more?" he asked in 1860. He wondered why people did not donate land to the town. He saw that trees—as well as hills and ponds and natural landmarks—were as much a part of the town as its homes and statuary, and that the landscape had a civic value, a value off the books. "What are the natural features which make a township handsome?" he asked his journal the year before he died. "A river, with its waterfalls and meadows, a lake, a hill, a cliff, or individual rocks, a forest, and ancient trees standing singly. Such things are beautiful; they have a high use which dollars and cents never represent. If the inhabitants of a town were wise, they would seek to preserve these things, though at a considerable expense; for such things educate far more than any hired teachers or preachers, or any at present recognized system of school education."

As many writers will tell you, it's when you are trying to concoct opinions and publish them, when you are thinking with the market in mind, that you run into difficulty publishing. When you are not so much thinking but doing, when you are writing because you see no other alternative, things tend to go much more smoothly, even though it can take a lifetime of work to get to that point. Thoreau was never a fan of politics; he was the last person who would have run for office and was appointed to the positions he held. Yet he became more engaged with the issue of slavery as a public speaker after he left Walden. He found himself writing (and subsequently delivering and publishing) a few brilliant and almost effortless-seeming public speeches. Slowly, he had become the writer's version of the civil citizen he had hoped to wake up with *Walden,* standing up, acting as a friction in a nation of yes-men.

In his private life, he was already engaged in the slavery debate, of course. Aside from their meetings and protesting, his family aided fugitive slaves. In his journal in October 1851, he describes a fugitive slave, Henry Williams, who had been attempting to purchase his freedom from his master, who was also his father, but had not been able to raise the entire $600 and, upon hearing of warrants out for his arrest, fled to Concord, where he had been directed to the Thoreaus. "He lodged with us," Thoreau wrote, "and waited in the house till funds were collected with which to forward him." Thoreau was about to put him on the train to Canada, via Vermont, when he noticed a man who appeared to be an undercover Boston policeman: "Intended to dispatch him at noon through to Burlington, but when I went to buy his ticket, saw one at the depot who looked and behaved so much like a Boston policeman that I did not venture at that time." This is a Thoreau out of a 1970s Gene Hackman movie.

On July 4, 1854, shortly after *Walden* was published, Thoreau spoke at the Anti-Slavery Convention at Framingham, Massachusetts. The speech, "Slavery in Massachusetts," printed in William Lloyd Garrison's *Liberator* two weeks later, was a rabble-rouser. At the event, a copy of the Constitution was burned. If "Civil Disobedience" was a call for a practical response to slavery, a plan for resistance to government in contrast to nonresistance, then Thoreau's speech in Framingham proposed the next step, the position of someone a little less civil, even more thirsty for action. With the Fugitive Slave Act, the federal government, with the aid of the state and city governments, sent agents to Massachusetts to forcibly take black men and women back to the South. Right before Thoreau spoke at Framingham, a group of abolitionists had stormed a Boston jail to free a black man awaiting trial—a judge was about to decide whether the man was free or a slave. In the action, a guard was killed, and federal troops descended on Boston. For a few days, Boston was a military state, under martial law. Thoreau understood that the battle for slave states that some easterners considered a western dilemma had now come home to New England. Merchants could no longer look the other way, because the issue of slavery was in every view. As usual, Thoreau's concern was with his neighbors, the local populace, who are busying themselves with faraway concerns—whether to allow slaves in Nebraska, in other points west—and not seeing that the faraway concerns are in their midst. Thoreau's opening is still a powerful attack on smugness, on targeting problems far-off as opposed to those nearby:

> I lately attended a meeting of the citizens of Concord, expecting, as one among many, to speak on the subject of slavery in Massachusetts; but I was surprised and disappointed to find that what had called my townsmen together was the destiny of Nebraska, and not of Massachusetts, and that what I had to say would be entirely out of order. I had thought that the house was on fire, and not the prairie; but though several of the citizens of Massachusetts are now in prison for attempting to rescue a slave from her own clutches, not one of the speakers at that meeting expressed regret for it, not one even referred to it. It was only the disposition of some wild lands a thousand miles off which appeared to concern them. The inhabitants of Concord are not prepared to stand by one of their own bridges, but talk only of taking up a position on the highlands beyond the Yellowstone River. Our Buttricks and Davises and Hosmers are retreating thither, and I fear that they will leave no Lexington Common between them and the enemy. There is not one slave in Nebraska; there are perhaps a million slaves in Massachusetts.

Another way to encourage your neighbors not to remember you very fondly is to get up and give a speech that says they are all sitting on their hands.

At the end of 1859, as I have previously noted, he came out again to speak following the death of John Brown, this speech even more radical than the first. Thoreau rushed out this speech, sending a boy around Concord to tell people that he was to speak on the subject. Word came back from town leaders that they thought it was not a propitious time, that a respectful wait might be in order. Thoreau sent back word that he had not asked them whether or not he should speak but was merely telling him that he was going to. He rang the town bell himself. If he did not take up arms in his life (despite hunting and, for a time, his requisite training marches with the local militia), he nonetheless became more violent in his political speeches. Thoreau is clearly no nonresister. Recall from "Civil Disobedience": "I do not wish to kill or be killed, but I can foresee circumstances in which both these things would be by me unavoidable." Now he accepted John Brown's violence as inevitable. "It was his peculiar doctrine that a man has a perfect right to interfere by force with the slaveholder, in order to rescue a slave," he said. "I agree with him."

Thoreau may or may not have known that Brown had brutally slaughtered a farmer and his son, among others. The newspapers were focused on strategy, on whether Brown's raid was or was not a success. Thoreau was focused on Brown's righteousness, rather than the rights of those citizens who agreed with slavery. Didn't Brown throw his life away resisting the government? What would he have gained by it? These were the questions in the air. Thoreau's response: "Well, no, I don't suppose he could get four-and-sixpence a day for being hung, take the year round; but then he stands a chance to save a considerable part of his soul—and such a soul!—when you do not. No doubt you can get more in your market for a quart of milk than for a quart of blood, but that is not the market that heroes carry their blood to." The writer Lewis Hyde has noted that while Harriet Beecher Stowe wanted us to feel and pray, Thoreau wanted us to act.

IF YOU JUDGE ONLY by what is in Thoreau's anthologies, it can even seem as though Thoreau gave up nature for rabble-rousing speeches about politics in his final years, but for Thoreau, of course, nature and society were always linked. In "Slavery in Massachusetts," he wrote, "I walk toward one of our ponds, but what significance the beauty of nature when men are base?" Nature is less beautiful when viewed by the tainted individual. Man can affect nature. Men do affect nature. People can make for beautiful change. For Thoreau, nothing separated the church and state of nature and man. Again in "Slavery in Massachusetts," he described the scent of a white water lily, pure and beautiful, growing in the "slime and muck of earth." A person should behave so that the odor of his actions "might enhance the general sweetness of the atmosphere, so that when we behold or scent a flower, we may not be reminded how inconsistent our deeds are with it." You smell the way you act, in other words. You are your odor. The lily smells sweet because of good deeds. "If fair actions had not been performed, the lily would not smell sweet," Thoreau said to the crowd on that Fourth of July. "The foul slime stands for the sloth and vice of man, the decay of humanity; the fragrant flower that springs from it, for the purity and courage which are immortal."

Whether he was talking about farmers or abolitionists, Thoreau was always endeavoring to connect man and nature. You can imagine him on all those long walks thinking about how nature and "the foul slime of man" work together, like compost and roots, to transform each other, to create something beautiful, a lily in the swamp. Decay leads to fruition, even in that slimiest of human endeavors, politics. The end is a beginning. In burying slavery, Americans would be adding compost to the national land:

> Slavery and servility have produced no sweet-scented flower annually, to charm the senses of men, for they have no

> real life: they are merely a decaying and a death, offensive to all healthy nostrils. We do not complain that they live, but that they do not get buried. Let the living bury them: even they are good for manure.

HE CAUGHT HIS LAST COLD IN THE WINTER OF 1861, from Bronson Alcott, who dropped by as usual with an idea, and this time left Thoreau with pneumonia. The cold turned into influenza, and the influenza turned into severe bronchitis, which kept him in the house all winter. In a daguerreotype taken shortly before he died, his eyes are sunken, tired, his face old beyond its forty-four years. He had been writing fewer pages of observation in his journal, but he kept combing through his notes and previous journal pages, collating his charts; he was not about to let up on his big projects. No one knows for certain what the big projects were. A book on the succession of seeds? A book about all the wild fruits he knew? His Concord calendar, the cultural landscape and interactions of all four seasons of his hometown, his forever explored pellicle of land?

When spring came, the doctor recommended he change climates for his health. Instead of Europe or the Caribbean, he decided on Minnesota, where he hoped to meet western Indians. His regular traveling partners were not able to go with him, so he went with Horace Mann, Jr., the seventeen-year-old son of the education reformer. Minnesota was still something of a frontier then, the edge of the overly described world, and fertile ground for those with interests in botany and Native American culture. As far as Indians went, he saw what must have been a depressing scene for him: a tribe of Chippewa who danced briefly before receiving checks from the government. (The following year the same tribe would stage an armed uprising against the troops guarding their reservation.) His experiences with Indians this time were completely different from the experiences he had had in Canada with Joe Polis, the guide he had hired there and profiled in his reporting on the Canadian woods; Polis had shared language with him and had impressed him with his ways in the woods, the kind of guy who appealed to Thoreau's romantic conceptions of Native Americans in general. Thoreau wrote a few short letters home, but his must have been a daily battle; people he met described him as struggling to settle his chest each time he spoke.*

Thoreau began to wrap things up in January of 1862. James Fields, the latest editor of the *Atlantic Monthly,* had written to ask him about contributing to the magazine. Thoreau had not forgotten how the previous editor, Lowell, had changed his text without asking. "Of course, I should expect that no sentence or sentiment be altered or omitted without my consent," Thoreau wrote. Fields returned with a very good price per word, and Thoreau began pulling together his old lectures to run as essays—a good price per word will even move a writer on his deathbed. He sent "Autumnal Tints," "Life Without Principle," and "Walking." A week before he died, he was trying to straighten out the last portions of his Maine woods writing: "It is in a knot I cannot

* While he was in Milwaukee, Thoreau, by coincidence, appears to have come very close to crossing paths with John Muir, a Wisconsin native who would later take to heart Thoreau's writings on the importance of land conservation. The relationship between Thoreau and Muir is nicely encapsulated by the poet Robert Haas, who adds the two men into a longer chain that combines poetry and politics: William Wordsworth is read by Henry Thoreau. Thoreau is read by John Muir. Muir is read by Teddy Roosevelt. Roosevelt writes the bills that make national parks: poetry in action.

untie." Fields offered to reprint *Walden,* and even came to Concord to buy up all the old copies of *A Week* that had lived in the Thoreaus' attic for so many years. The last essay Thoreau sent him was "Wild Apples." It had been his last lecture in Concord, culled from a longer manuscript, called "Wild Fruits." For a long time, scholars lumped "Wild Apples" into a pile of late writings that seemed less significant than *Walden* or his political writings. For many years it appeared to be merely about apples, and of course it is. But it is also a resolution of the difficulties he had with lecturing and publishing in general, with his writing life. It's an addendum to *Walden* by the artist as an old man, only boiled down to a few pages, like sap.

IT WAS A FAMILIAR ESSAY, the kind you might read in a magazine today—the author explaining his interest in apples in general and wild apples in particular. It was an easy topic: everybody loves apples. All through his career he had sought to bridge the gap between facts and truths. Lecture audiences wanted facts and yucks, while he wanted so badly to give them insights and truths. To the end, Thoreau wanted to deliver a lecture that would please the "hasty and deliberate" audience as well as the Transcendentalist. His first book had been overloaded with truths, to its critical detriment: "We were bid to a river party, not to be preached at," wrote James Russell Lowell, his editor turned editorial archenemy. The apple is as literal as it is symbolic, the source of cider and the fall of man.

He traced the history of the apple tree through Roman literature, through European sources, following it on a western migration, and in so doing, quietly questions the idea of progress: "It has been longer cultivated than any other, and so is more humanized; and who knows but, like the dog, it will at length be no longer traceable to its wild original?" The apple orchard becomes not the place where man takes on sin, but the place where he can renew his innocence, where he can recover his wildness "without price, and without robbing anybody." The cultivated stock becomes the European settler, the wild apple the man who has adapted to nature, the crab apple the Native American—Thoreau had respect for the crab apple, and did not want to see it cultivated but wanted it studied, appreciated, understood. As usual, Emerson hated that Thoreau liked what everyone else considered weeds.

"Wild Apples" is also personal biography, about as personal as Thoreau got in his published work. You can hear him describe his homeliness, in a good way: "Almost all wild apples are handsome. They cannot be too gnarly and crabbed and rusty to look at. The gnarliest will have some redeeming traits even to the eye." The section entitled "How the Wild Apple Grows" seems to address his writerly frustrations. He describes the cows ("a fellow emigrant from the old country") as they browse on the shoots of the tree. The tree is cut back year after year. The tree does not despair, though. It forms a mass of sticks and stubble, of knotted thorniness that is, Thoreau says, "almost as solid and impenetrable as a rock." Decades can pass until at last, a shoot emerges, an upward-reaching branch that the cows can't eat, and the new tree is born. "For it has not forgotten its high calling, and bears its own peculiar fruit in triumph," he writes.

Even as a writer who struggled for acceptance, who had difficulty getting published by his own friends, who watched contemporaries like Hawthorne and, of course, Emerson grow into fame and commercial success in publishing, he was astoundingly hopeful that there would be fruits to his labors, if only in the future—a future he knew he would not live to see.* The dung could be scientifically said to serve the future of poetry, of dung-buried literary treasure. All the years of dung would serve the wild apple in the end—a take-home point for writers even nowadays. You can read "Wild Apples" today as an homage to locally grown fruit, as a treatise on the benefits of indigenous species. Or you can read it as the assertion of the value of local artistry, as the pronouncements of the at-last-matured American Scholar that Emerson had called for at the beginning of Thoreau's career. Now the scholar is hidden in the brambles, in a secret, no-longer-populated dell: "Do not think that the fruits of New England are mean and insignificant while those of some foreign land are noble and memorable." You can also read "Wild Apples" as an appropriatory quiet meditation on the value of working quietly, on persevering. Any way you look at it, however, you can feel that the author is ultimately sad to be leaving Concord at last and forever, his long walks winding up:

> The note of the chickadee sounds now more distinct, as you wander amid the old trees, and the autumnal dandelion is half-closed and tearful.

* The essay is a little less bitter than the journal entry, its cuts hidden. In his journal, he wrote about a man worn down with difficulties. "But at length, thanks to his rude culture, he attains to his full stature, and every vestige of the thorny hedge, which clung to his youth disappears, and he bears golden crops of Porters or Baldwins, whose fame will spread throughout the orchards for generations to come, while that thrifty orchard tree which was his competitor will, perchance, have long since ceased to bear its engrafted fruit and decayed."

When he was about to die, when tuberculosis finally had him pinned down, everyone knew it. A poet from Leedsville, New York, who had read Thoreau's books but never met him, wrote him to say that news of the author's illness had affected him as if it had been news of the illness of an old friend. "I am encouraged to know, that, so far as you are concerned, I have not written my books in vain," Thoreau wrote, adding, "I *suppose* that I have not many months to live." The Civil War had broken out, the bloodshed he had imagined in "A Plea for John Brown," and even "Civil Disobedience." His illness had been reported in the papers, as his death would be. "King of the Gypsies" the *New York Times* called him, writing, "He was on talking terms with oaks. The aspen forgot to tremble in his presence, the mimosa to shrink at his approach." Already posterity was calling him Nature.

His family moved his bed down to the parlor, where people stopped for palaver—noting that he was an inspired speaker in the last weeks, that, although his voice was no longer stronger than a whisper, he was talking a blue streak, pausing only to catch his last breaths. His sister now took dictation. There was a steady stream of visitors. His friend H. G. O. Blake ice-skated down from Worcester to see him, Thoreau saying that he had not arrived too soon. Blake described him as being in "an exalted state of mind" as he spoke of skating on some other river. "Perhaps I am going up-country," Thoreau told them. Sam Staples, the man who had jailed Thoreau, dropped by. Emerson stopped by repeatedly, talking about the river, birds, ice. "Never saw a man dying with so much pleasure and peace," Emerson wrote. Sam Staples concurred. It was the opposite of his brother's grueling, excruciatingly unpleasant death, a peaceful passing. Thoreau was pleased with all the visits, proving again the vanity that Hawthorne had noted. "I should be ashamed to stay in this world after so much had been done for me," he told his sister. "I could never repay my friends."

He was distressed that he could not venture outside, naturally. Scraping frost off the window, he said, "I cannot even *see* outdoors." He heard an organ-grinder playing a tune from his childhood and forced a family member to run out and pay the guy. He wrote a journal entry about a cat playing indoors, which when you read it is just a little pathetic for the great taker of long, purposefully aimless walks. Friends who were more orthodox in their religious pursuits were upset that he did not want to prepare for death in traditional ways. He told one of them that a snowstorm meant more to him than Christ. His friend Parker Pillsbury, an old abolitionist and minister, asked him if he could see across to another world. "One world at a time," Thoreau replied.

When Sophia read him the Thursday section of *A Week on the Concord and Merrimack Rivers,* he said, "Now, comes smooth sailing," a reference to that last day of the trip he took with his brother, when, in a stroke of luck, they caught a wind all the way home to Concord, comfortably gliding down the river toward home. His last sentences were difficult to understand but are often thought to have the words *moose* and *Indian* in them. This has elicited much speculation among academics and Thoreauvians, but it seems clear to this professional free-lance that he was wrestling with his Maine papers, trying to get them into shape for publication, trying to make money for his family as he left them, and anxious to the end about meeting a deadline.

After he died, on May 6, his body was brought to church. Wildflowers were laid across the coffin. Emerson delivered the eulogy. Alcott let the children at his school out early. "Though he wasn't made much of while living," Louisa May Alcott noted, "he was honored at his death."

Chapter 12

PERFECTLY DISTINCT

I WENT UP TO SEE WALDEN POND not too long ago, driving up from Brooklyn, where I live. I had been to Walden only once before—a quick trip with my wife and children, at the end of a long round-trip drive across the country that had made us eager to get home. All through my time writing this book, I resisted visiting Walden; I just wanted to say a few things about Thoreau, which, from the vantage point of the ending, I can see dragged into too many things, I'm afraid. Still, I wanted to visit the site of Thoreau's house once before I closed the book on him—or closed this one, anyway. So last spring, I came in off the interstate and caught State Route 2, which was heavily trafficked and semi-interstate-ish at times. Fighting oncoming traffic, confused about unfamiliar turn patterns, I took a slightly desperate left into

Concord, not realizing I had been pretty close to Walden Pond while on the highway. The traffic quickly thinned, except for the guy who was on my tail driving into downtown, who made me anxious and kept me from fully enjoying the scenery. Eventually, I pulled over in front of the library to let him pass. I took a deep breath. We looked at each other as he drove by.

I checked my maps, keeping my finger on the hotel I was planning to stay at while I pulled back into traffic, moving carefully around the town square, where I easily spotted the Colonial Inn in a building once owned by Thoreau's grandfather, John, the pirate, who came to America when shipwrecked near Boston. The Colonial Inn was a little pricey in comparison to the motel out on the highway, but the staff was exceptionally nice, and I wanted to do this Concord visit right, even luxuriously, so I threw down my credit card, gulped, and checked into a small historic room with a big TV. Then, I dumped my car in the big parking lot out back and set out to walk to Walden Pond.

I stopped first at the local post office, to buy some postcard stamps. I like post offices. Thoreau talks about meeting people at the Concord post office. At the Concord post office, an old man once told him that he considered Thoreau and Emerson and Channing to be a kind of club, the Walden Pond Walking Club, the man called it. Thoreau was pleased to hear this. I would imagine that some people in town thought the club members were nuts, while some people thought it was great that they were out walking so much. Some people might have thought, Don't they have any work to do? Some might have thought, I've got to get out walking in the woods once in a while. And probably some people probably didn't think about it at all. As I paid for postcard stamps, I asked if I could get to Walden Pond from there by heading down Walden Street, which seemed right, according to my maps.

"Yes, sure," one of the post office workers said to me. "Where is your car parked?"

I told him that I planned to walk to Walden Pond.

"You're *walking* to Walden Pond?" he asked incredulously. Suddenly the people in the other two lines were looking at me.

"Yeah," I said. "I've got some time, and I thought I'd walk."

One of the people in line next to me asked the person next to her, "Can you *walk* to Walden Pond?"

"You drive," the woman said.

The postal clerk I was talking to was helpful and patient with my situation. "Well, it's a good walk, but yeah, that'll get you there," he said. "Just go straight down, I guess."

FEELING A LITTLE LIKE A FREAK, I SET OUT—first down Walden Street, then via Hubbard Street to Lexington Road, where the traffic picked up, but I managed to sprint across the street to the Concord Museum. Inside, I watched a nice slide presentation. I read about Concord as a popular suburb of Boston and saw a quote from Hawthorne that seemed to describe a very different Concord from that of Thoreau, a docile, picturesque landscape as opposed to a bustling, working one: "The scenery of Concord, as I beheld it from the summit of the hill, has no very marked characteristics, but has a great deal of quiet beauty, in keeping with the river." I bought a card with one of Thoreau's quotes on it; it was from "Huckleberries," a late essay. I planned a note in my head and thought of whom I would mail it to, though when I got home, I never got around to it—it's staring at me on my desk right now:

> Live in each season as it passes; breathe the air, drink the drink, taste the fruit, and resign yourself to the influences of each.

That's one of those Thoreau quotes that can be considered either corny, or absolutely true, and even though I had just gotten out of bad traffic and paid more than I could probably afford for a hotel room, it was feeling true to me, as I was in a Thoreauvian mood, ready—maybe even *too* ready—to have some great Walden Pond experience, if possible.

I looked at the museum's replica of the house Thoreau built at the pond. No one else was looking at the house, and it was locked, or seemed to be. Across the street from the museum was Emerson's house; it was closed to visitors and set up for what looked like a private party. Concerned about the distance after talking to folks at the post office, I got back on the road to the pond. It was mid-afternoon, and no one was walking. Just a couple of blocks from the center of Concord, I already felt all alone, cars whizzing by. Maybe it would be different if I lived there, but right then I felt the isolation that a lost tourist feels. I felt separate.

IN THE YEARS AFTER THOREAU'S DEATH, his reputation was a little like the house he built at Walden. It had a high profile locally for a short time, and then slowly faded into the woodwork—literally faded into the woodwork, as the walls became part of the walls of a barn: a Thoreauvian metaphor made real. Newspapers noted Thoreau's death and his last published pieces, especially the essays for the *Atlantic*

Monthly on Maine. His last two books, *Cape Cod* and *The Maine Woods,* were published. But then he slipped into the shadows, primarily as the lesser Emerson, the eccentric nature poet. Emerson—unintentionally, it seems to me—helped in this regard, starting with Thoreau's eulogy, also published in the *Atlantic.* Aside from being what biographer Robert Richardson has called "the best single piece yet written on Thoreau," it is also the piece that set him up as a "bachelor of thought and nature," as a "hermit"—i.e., the freak alone along the road.

Thoreau's family denied that their son and brother was a hermit, and argued about Emerson's essay when it was published, but the charges stuck. Emerson was the sage, Thoreau the village character: a good guy, basically, a smart guy, but a little too much. With Emerson as a literary friend, he didn't really need a literary enemy, but he had an excellent one: James Russell Lowell, his old *Atlantic* editor, another literary hotshot who was very intentionally against him. Lowell reviewed his previous anti-Thoreau musings in his memoir, *My Study Windows;* he charged Thoreau with being overly egotistic, with a "morbid self consciousness that pronounces the world of men empty and worthless before trying it." Lowell further accused Thoreau of ignoring the abolitionist movement altogether, which probably also irritated Thoreau's family, who, as noted, were ardent abolitionists. Perhaps most damningly, Lowell accused Thoreau of having no sense of humor, a charge that, even though it was not true, obviously had some legs: it's one of those things that you say about somebody that really stings and subsequently sticks.

Even Thoreau's old friends mangled his legacy. In *Thoreau, the Poet Naturalist,* one of the first books on Thoreau, Ellery

Channing downplayed *Walden* and all of Thoreau's political writings. Harrison G. O. Blake was Thoreau's longtime philosophical pen pal. He was willed Thoreau's journals by Sophia Thoreau, all two million words of them. Despite years of letters from the author himself that talked about how to live in society and how to live a moral life, despite scores of letters that were less like nature poems than friendly sermons from a minister of an American church-at-large—despite all that Blake began publishing edited selections from Thoreau's journal in 1881 and only chose sections concerning nature. "Early Spring in Massachusetts," "Summer," "Winter," and "Autumn" are section titles. Meanwhile, in Concord, Emerson continued to be exalted above all others. In Oliver Wendell Holmes's 1885 biography of Emerson, Thoreau was characterized as a prankster: "half college graduate, half Algonquin, the Robinson Crusoe of Walden Pond, who carried out a school-boy whim to its full proportions, and told the story of Nature in undress as only one who had hidden in her bedroom could have told it." (It's a slam that actually reads pretty nicely, even after all these years.)

Thoreau's publisher, Ticknor and Fields, pushed him into the literary marketplace. That his journals were published at all—he was the first American author to have his journals published—was an indication of how hard they were pushed. Things began to turn for his reputation when, in 1889, Emerson's son, Edward Emerson, wrote a reminiscence of his father, *Emerson in Concord,* that made Thoreau seem more human. (Edward Emerson's portrait of Thoreau would come out in 1917, and he would remember him as a kind uncle, who took the kids into the woods, who was good with practical matters, especially relating to the Thoreaus' pencil factory.) Around the same time, naturalist John Burroughs rebutted the idea of Thoreau as a humorless "skulker," and he refuted Emerson's charge that Thoreau was limiting himself as "captain of a huckleberry-party." But Burroughs otherwise confirmed Emerson's appraisal: "He was, by nature, of the Opposition; he had a constitutional No in him that could not be tortured into Yes." Burroughs chided Thoreau's trip to jail, calling it "uncompromising" and "almost heartless." Thoreau came out a good naturalist, *à la* Burroughs, but a crank, as opposed to Burroughs: Thoreau's love of nature was to be commended but his critique of society was over-the-top. In a way, Burroughs used Thoreau's writing to affirm the job of the naturalist—Burrough's own job—at a time when the public's nature-loving was moving into high gear in the culture and when he was ideally situated as the man to shift it higher. To link politics to nature, to link ethics to the poetry of the outdoors, would in a way spoil the vacations of the class of Americans who could afford to buy homes in the hills, who were visiting the vistas of Hudson River painters. The politics of nature was still linked to politics primarily via the politics of empire. Burroughs didn't see that by saying "no" to the demands of the market and society, Thoreau was saying "yes" to everything else.

Thoreau had a more positive reception in England upon the publication of *Walden*. The English audience more readily understood it as a social critique. George Eliot—who called *Walden* "a bit of pure American life (not the go-ahead species, but its opposite pole)"—criticized the American critics: "People—very wise in their own eyes—who would have every man's life ordered according to a particular pattern, and who are intolerant of every existence the utility of which is not palpable to them, may pooh-pooh Mr. Thoreau and this episode in his history, as unpracticable and dreamy." If the Americans took him literally, as a guy in a cabin, the British were able to see him as he hoped (and punned)—as the extra vagrant. In 1877, the British author A. H. Japp, in *Thoreau: His Life and Aims,* wrote that Thoreau was criticizing "the artificial make believes of modern society." Japp saw Thoreau not as a misanthrope but as someone who "loved individual men, and most that which was individual in them." More importantly, Japp saw a link between the ideas of nature and society. Thoreau's love of nature, he wrote, "did not lead him to sour retreat from society, but rather to seek a new point of relation to it, by which a return might be possible and profitable." Even though he got Thoreau right, Japp's *Thoreau* was, unfortunately, a hagiography. It was panned by critics.* It was a way of thinking about Thoreau that would not be picked up again for a long time.

BACK ON WALDEN ROAD, OR, MORE PRECISELY, looking for Walden Road, I went south for a few yards on Heywood Street, a short street that connected Walden Road and Lexington Road, and in so doing, I passed what looked like an old orchard adjoining the Emerson property. *Imagine*

* A comparison in the book that unsettled critics was Japp's comparison of Thoreau to Saint Francis. This comparison is interesting to note, in that the Catholic Church, and people in general, tend to emphasize Saint Francis's relationship with the natural world; on the feast day of Saint Francis, Catholic and Protestant churches traditionally bless pets and animals, for instance. The historical person of Francis is more Thoreauvian than Japp's detractors might have acknowledged. The son of a wealthy Umbrian landowner, he renounced his wealth, and lived in self-imposed poverty. He lived and worked (and arranged that his followers might live and work) with those whose poverty was not self-imposed. He proposed negotiation during the Crusades, rather than force, and was said to be admired by Muslim leaders, though they rebuffed his arguments for converting.

Thoreau picking apples, I thought to myself. I also imagined Thoreau thinking, *Why doesn't Emerson ever pick his own goddamned apples!* I saw a pleasing open space, a field and trees. But I didn't see any people, other than drivers; the road to Walden Pond felt very back-road that day. In a minute, it felt very edge-of-town. The speed limit was posted at twenty-five miles per hour, but cars were going faster—a really nice Ford Mustang, for instance (one of two Mustangs I would see that day). Old houses that looked like the homes of noteworthy Civil War–era Concordians disappeared and newer housing replaced them, little houses with small parking lots in front, and, at one of these houses, three people standing around their car said hello when I waved hello to them. Then, just before a bend in the road the sidewalk disappeared. I began to walk less leisurely, repeatedly looking back over my shoulder. A cop car passed, after which I came upon Concord's public safety complex, a fire and police station—did they have a jail? When a sentence sprouted in my head about how bored the police and fire cars appeared, I thought again of Emerson, who would have told Thoreau that police cars can't be bored—the kind of comment that would have driven me, personally, nuts.

The road seemed to narrow. The speed limit increased to thirty-five miles per hour; cars seemed less enthusiastic about my presence, if they noticed me. In a minute, I saw what appeared to be a large community garden, two hockey sticks marking a plot of space, a regional touch. No one was there. It was a weekday, around three, and everyone was off at work, making a living.

Just then, my cell phone rang. I was trying to not take any calls while walking to Walden Pond, but it was my home phone calling. My wife asked me where I was.

"I'm walking to Walden Pond," I said.

"Okay," she said, "be careful."

I could hear another car coming. I had a vision of being plowed down while on the phone. I stepped in from the road, and then talked to my son, who was the reason for the call; he was looking for the phone number of a friend of mine. My friend was out of town.

"He doesn't have a *cell phone*?" my son said.

"No," I said. "He doesn't."

THE RESETTING OF THE IMAGE OF THOREAU as an antitechnology, anachronistic nature worshipper was begun slowly in the 1890s by a small group of correspondents in America and England. Samuel Jones was a homeopathic physician who lived in Ann Arbor, Michigan, and, in his spare time, studied Carlyle and the American Transcendentalists, especially Thoreau; he had begun a correspondence with A. W. Hosmer, a photographer in Concord who was a relative of Edmund Hosmer, the neighbor whom Thoreau had called the "long headed farmer,"—which, in the phrenological terms popular at the time, denoted a certain shrewdness on the farmer's part. Henry Salt was a social reformer who was born in India but grew up in England and came upon Thoreau through the early Victorian socialists, in particular Edward Carpenter, who was, among other things, a staunch proponent of the simple life, a philosopher, and an early gay activist. (Carpenter's book, *Civilization: Its Cause and Cure,* argued that civilization, rather than purely beneficial, is a kind of disease that man passes through and typically lasts about a thousand years.) It was Salt's biography of Thoreau, which appeared in 1890, that was eventually read by Gandhi, and it was through Salt that Thoreau's writing was carried around by early members of the British labor party.

The writers corresponded on whether or not the shack at Walden was a station on the Underground Railroad, as had been stated (it was not), and over who paid his tax to free him from jail. "I had got my boots off and was sittin' by the fire when my daughter told me, and I wasn't goin' to take the trouble to unlock after I'd got the boys all fixed for the night, so I kep' him in 'til after breakfast next mornin' and then I let him go" is the comment they got from Sam Staples, the jailer. They tried, according to Jones, "to redeem Thoreau's character from the reproach of being so cynical that he hardly seems to have belonged to the HUMAN race." Jones spoke for all three Thoreau lovers when he wrote: "In 1891, Emerson's false picture has not been corrected; Thoreau is to the majority a 'stoic,' but Edward Hoar [a neighbor of Thoreau's] can scarcely speak of anything other than Thoreau's infinite tenderness." They collected tidbits—letters, relics, anecdotes about, for instance, Thoreau showing snakes to kids. They criticized Thoreau's critics. "Torrey is of the earth earthy, and I much question if Thoreau would have admitted him to one of his 'walks,'" one fan wrote to another. They quarreled, especially with Frederick B. Sanborn, a Concordian who seems to have tried to corner the market on Thoreau. "Draft F.B.S.," Jones wrote. "I'd like to pull his nose!"

They were particularly obsessed with Thoreau's relationship with Ellen Sewall, who had rejected both Thoreau and his brother. They seemed to think that it would prove Thoreau human, that a lover—and in particular a lover of the opposite sex—would reveal what Jones called "a side that [Thoreau] kept sedulously hid from the world." They agreed that the marriage of Thoreau probably had to do with religious problems. The *not* marriage. They talked about Henry "in LOVE," as if competing with the comment that Emerson quoted in his eulogy for Thoreau: "'I love Henry,' said one of his friends, 'but I cannot like him; and as for taking his arm, I should as soon think of taking the arm of an elm-tree.'"

Thoreau's bachelorhood was problematic for them, as it began to seem problematic for Emerson—as it was for newspapers everywhere. In the early reviews he was repeatedly called "eccentric." *Putnam*'s called him a "lusus," a sport. The *New York Times* mentioned his "eccentricities." The *Yankee Blade* described him as having "an odd twist in his brains," while the *Alta California* recommended *Walden* to "all fops, male and female." Eccentric was often code for what we now call gay, and a lot of the same papers derided the eccentricity as "selfishness." *Graham*'s said: "He differs from all mankind with wonderful composure; and, without any of the fuss of the come-outers, goes beyond them in asserting the autocracy of the individual." Emerson, in his eulogy, can seem to attempt to paper over any fears people might have about Thoreau's eccentricity—he compared him to a noble youth in search of a particular rare flower that fair maidens in particular love. It's as if Emerson was worried about protecting him, which is too bad. Better to be a noble hermit than a selfish eccentric, better a high priest than a "nature boy."*

* As the selfish nature boy, as the eccentric unmarried man, even as the crank, Thoreau ends up being less worth listening to when he is arguing about the state of our state. When he speaks of the exchange of citizenship for consumerhood, we can respond by saying that he wasn't married and lived alone—that he didn't know how to be social, how to live a noneccentric—or, *normal*—life. Emerson was working on, in the words of the scholar Henry Abelove, "his domestication of Thoreau." Abelove sees *Walden,* on the other hand, as the "the first queer action," a political act in that it refuses domestication, a protest of

---

STILL NO SIDEWALK ON WALDEN ROAD, and though I was less than a half a mile from town, I felt, mostly because I was walking, as if I were a fairly well-dressed tramp. At some point, a little dirt trail on the road's shoulder appeared. On my left, I saw an office building of some kind that was so without distinguishing detail, and so alone in the roadside clearing, that I could not for the life of me guess what it was. On my right, I saw a faraway complex of buildings, off through the cleared fields. A loudspeaker mumbled in the distance, and I guessed (correctly) that it was coming from Concord's elementary school, named, as I later learned, for Bronson Alcott. The open area of cleared fields reminded me of what the area would have looked like when Thoreau was a kid, when it was all open, as opposed to now, when it was mostly woods, cut down only for new shopping or housing developments.

Just about at the halfway point to Walden, I came to a bend in the road that forced me to pick up my pace. The little dirt path all but disappeared; the hedges pushed in from the shoulder. Because I was anxious not to get run down by a car, I crossed the street, to walk with oncoming traffic facing me. I jogged through the turn, only once pressing into the bushes as a car raced past. *I'm still alive!* I said to myself, in the manner of either a chanticleer or a chicken. On my left were woods—to where, I do not know. On my right was a

---

mores. The narrator of *Walden* went to town, according to *Abelove*, "to see the men and the boys." He reads the scenes with the woodchopper as a representation of seduction. Whether you think of Thoreau as gay or not, I think it's important to recognize that Thoreau is trying to cherish, preserve, and promote ways of life—ways of being, even—that are being weeded out, and that criticism of him for his otherness—criticism of his person, in other words—is often just a way of ignoring his greater critique of society.

dwelling that I first thought was a quaint and nicely kept little house built for people who were able to live on a small scale. I was excited. A Thoreau-esque mini-manor! As I got closer, and as cars used the circular driveway to turn around and head back into town, I realized it was not a home but a house built for the phone company. The building was merely built to resemble a house. I enjoyed thinking that while I stood there quietly, the interior of the house was filled with electronic chatter, people talking or fighting or announcing they must have wrong numbers. I thought of the joke in *Walden:* "the man who was earnest to be introduced to a distinguished deaf woman, but when he was presented, and one end of her ear trumpet was put into his hand, had nothing to say." I thought of one of my favorite little poems by Thoreau, featured on the walls of the New York subway trains all through the time I was typing up this book:

Men say they know many things;
But lo! they have taken wings—
The arts and sciences,
And a thousand appliances;
The wind that blows
Is all that any body knows.

I instinctively felt my pocket for the vibration of my cell phone, as I often do, out of habit, a buzz in my pocket that's more like Pavlov's shock than music from Aeolus, if I may wax a little *Walden*-y. There was no call, naturally, no transmission, no incoming communication—it was an illusion.

STILL ON THE ROAD TO WALDEN, just past Thoreau Street, which led back to Concord—a sign at the intersection said, SHOPPING DISTRICT. Next, I saw signs announcing construction at the entrance to the local high school—photos and plans suggesting a new addition, a sports complex that I later learned was controversial among the people of Concord—trees in an area known as the Deep Cut Woods were cut as part of the project.* Deep Cut was a woodland area named by Thoreau for the railroad cut—the site of his sand-cut description.

After the school entrance, I proceeded up a small hill and saw a few more houses on the right, and then a little road. The sun had broken through a haze now, and the woods behind the little houses went from being dark to a bright kaleidoscopic green. Still no sidewalk, but a relatively untraveled shoulder and numerous No Parking signs—directed at high school students, I imagine.

And then I arrived at the intersection of Walden Street and Route 2, a major crossing, as in *major*. You could say I am exaggerating, but it's a highway, and I was dreading crossing, though perhaps it would not startle a Californian—or any American, come to think of it, who has made the long, expeditionary trek from a strip mall on one side of a highway to a strip mall on the other. I suppose that if you lived in Concord you would think nothing of it, as I would be less concerned by this crossing on my return: familiarity breeds the sensation of security. But if this was a story of a

* The Thoreau Society's Web page said: "Efforts to save Deep Cut Woods were not successful, and in July a crew began cutting down trees to create multi-sport fields for Concord-Carlisle High School and Concord-Carlisle Youth Sports. We thank everyone who supported the Deep Cut Woods cause." Deep Cut is where Thoreau did his extensive studies on the succession of trees, among other things. Note that the area was named for a man-made construction. Today we would be more likely to name a man-made area, like a housing development, for a non-man-made one: for example, The Mall at Walden Woods.

journey—which it is—then this would be the part where I wonder if I will even make it, the dark part, even though, after a mad dash across a single turning lane, I was now standing in the bright, sun-burning sunlight of the treeless traffic island, in the center of the traffic-loud glade. I looked for a button to press to initiate a light change. The sign above the button said: PUSH BUTTON FOR WALK SIGNAL. Above that sign was another that offered inherently counterintuitive advice: WATCH FOR TURNING VEHICLES ON WALK SIGNAL.

When you stand out alone on a traffic island alongside a four-lane highway, you feel like a character in a Samuel Beckett play—left alone, adrift, to wrestle with the loneliness that sometimes seems to be the main point of our race-like modern lives. And of course it's not as if you are wrestling with this loneliness while alongside a pond you can bide the time fishing in; you are wrestling with it while trucks whiz by, with a lot of honking, with exhaust. A car approached the island and honked at somebody or something and then, making a hasty right on red, accelerated madly. Among the many questions that I pondered concerning landscape and modern ways of life and the separation of nature and man was this one: I wonder if this button for the WALK signal is even working?

Suddenly, a man running for exercise joined me on the traffic island. He did not greet me; he had headphones on and did not seem to hear me say hello. I didn't take this as any kind of an affront; he was giving off a very positive vibe, just jogging in place alongside me. He wasn't ignoring me; we live in a culture where people wearing headphones—or earbuds, as these particular listening devices are referred to at this moment in technological history—or talking on telephones are not expected to greet the other person, and in many ways I am cool with that.* But feeling desperate about the crossing, I said hello again, and made hand motions that signaled I had a question. He graciously took his earbuds out of his ears, so that I could hear the music. I apologized for this inconvenience I was causing him. "No, no, no problem," he said, and he made me feel as if he meant it. "It's broken," he said. I thanked him. He put the earbuds back in.

This time, when the light changed and the traffic stopped for a second, the runner sprinted across the street, leaving me alone again. I was going to follow him but the traffic started right up. A minute later, I crossed the intersection of Route 2 and Walden Road. For those seven or eight seconds, I had that completely exposed feeling that everyone who has crossed a highway knows. I felt like a wild animal, like some woodchuck that was about to get flattened.

MOMENTARILY—IF YOU WILL ALLOW ME—I was but a metaphor for all the creatures of our world. Like three quarters of the land mass of the United States, the woods around Walden are touched by roads—in particular by Route 2, even if you'd never think about it while driving by or when hiking to the pond. Route 2 itself was studied in the 1990s by road ecologists at Harvard. They tried to determine what they called the "road-effect zone." They looked at bird nesting areas, deer migration routes, the effect of salt on the water along the road, the drainage of wetlands, the spread of invasive plants. In Concord, Route 2 passes over or near, among other things, a reservoir and two rivers, thirteen streams, and the

*On a brief fashion note, I would just like to add that being dressed in an Olympic-flavored running ensemble, he looked less out of place standing on a concrete island in traffic than I did in so-called street clothes.

airport that the Beatles landed at in the sixties. Moose come down occasionally from New Hampshire, moving back in on habitat after disappearing in Thoreau's time due to hunting. The reservoir had experienced high salt levels, and the streams appeared to have salt effects. Salamanders crossing from ponds to woodland areas were killed (though down the road in Amherst, the hometown of Emily Dickinson, tunnels were built beneath the road to spare the salamanders). Birds were affected mostly by the sound of the road. During incubation and the fledgling stages of reproduction, the sound of traffic is said to interfere with reproduction for the most sensitive of forest birds, up to two thousand feet away, which is about the distance from the corner I was on to the center of Walden Pond. From *Walden:*

> The Harivansa says, "An abode without birds is like a meat without seasoning." Such was not my abode, for I found myself suddenly neighbor to the birds; not by having imprisoned one, but having caged myself near them. I was not only nearer to some of those which commonly frequent the garden and the orchard, but to those smaller and more thrilling songsters of the forest which never, or rarely, serenade a villager—the wood thrush, the veery, the scarlet tanager, the field sparrow, the whip-poor-will, and many others.

Thoreau once said that when you cut trees, you were getting rid of birds. But the effect of the road on the woods is something else. Cutting a tree is, in many places in the United States today, extraordinary; driving is ordinary. Thoreau never said that roads would change the ecology of the United States and make disconnected islands of wildlife, plant and animal species that would be fundamentally less robust than if they were all connected. It was another writerly trope in Thoreau's day to be in the woods and describe the sound of the oncoming train, a knife in the heart of nature's solitude. But here traffic breaks into the woods without entering, an invisible interloper.

I WAS IN THE PLACE BETWEEN THE WORLD and Walden now, walking along a little path alongside a fence, on the other side of which I figured were Walden Woods—a distance of just a few feet gave me a feeling of protection. On my left was the line of cars waiting, as I had, for a signal to cross: the line was maybe a quarter of a mile long. The people in the cars either looked at me or didn't. It was a pleasant spring day, but almost all of their windows were closed. One car was thumping, the bass in the music shaking the windows, as if a party were trapped in the trunk. One guy had his window open, his arm and a cigarette dangling, singing a song louder than his car radio. A woman in a minivan looked at me as if I were nuts. Then again, who knows what she was thinking? Who knows what anyone is thinking?

Which reminds me of one of my favorite entries from Thoreau's journal, about a trip Thoreau took to West Acton, the town west of Concord. He and a friend were walking on the main road, the Harvard Turnpike. They came upon a large rock in the road, in the wheel track, the rut that wagons would have driven through. They noticed that the rock was unusually black. At first, they couldn't understand why. The rock was larger than a man can lift. They figured it had slipped off a wagon. That evening, at dusk, they were on their way home, coming down the same road.

> Returning the same way in the twilight, when we had got within four or five rods of this very spot, looking up, we saw a man in the field, three or four rods on one side of that spot, running off as fast as he could. By the time he had got out of sight over the hill it occurred to us that he was blasting rocks and had just touched one off; so, at the eleventh hour, we turned about and ran the other way, and when we had gone a few rods, off went two blasts, but fortunately none of the rocks struck us. Some time after we had passed we saw the men returning. They looked out for themselves, but for nobody else. This is the way they do things in West Acton.

AT THIS POINT, I ENTERED Walden Pond State Reservation, which is the official name of the state park that contains Walden Pond. I could see the pond through the trees. I did not pause but walked down toward the bathing beach. I recognized it as a moment of great drama, or potentially great drama, for me personally. Then again, what can you say about Walden Pond after you have spent a couple of years thinking about Thoreau? Try as you may to do otherwise, you often see what you see in part because of what other people see. Over the years, people have seen what they have wanted to see, and they mostly have wanted to see a kind of romantic beauty, nature unchanging, unspoiled, at equilibrium. In 1902, a dispatch from the pond and environs went like this: "Commercialism is a thing of which Concord does not boast." In 1929: "Much of the Pond's natural beauty is still present." In 1936: "Thoreau's cove, and the woods surrounding it, are still peaceful, and the lover of Thoreau will not find here too great interruption of his thoughts." (As if, with ice harvesting and timber cutting, they were nothing but peaceful when he was there!)

All this sameness, despite the changing, growing crowd: at the start of World War II, thirty thousand people were visiting Walden every year. In 1945, Edwin Way Teale, the naturalist, said that Walden was "wilder" even though he would have seen about ten thousand picnickers and swimmers a day—it was more wild if your idea of wild is bathing suits and sandwiches. *American Forests* sent a correspondent in 1969; he described Walden as "beautiful, wood-fringed, and as satisfying to escapees from today's urbanization as Henry David Thoreau described its environs nearly 125 years ago." In 1979, a reporter said, "the place seems pretty much as it must have been when Thoreau quit the cabin and went back to Concord."

I thought it looked good, I'll say that. Was the view special or meaningful? It was certainly meaningful in a way that has to do with it being a place where a lot of people have showed up over the years thinking it was meaningful. You might say Walden is littered with meaning. On this happy, summerlike late spring afternoon, there were not many people, as it was a weekday. On the beach I could see that a guy had been swimming. A little boy had been wading, and now his mother was explaining to him that he had to get the sand off his feet, and then scolding him. "No, honey! People don't want wet sand up there!" she said. A kayaker was paddling out toward the center of the pond, recreating, taking in nature as a form of exercise. I stood at the edge of the pond.

I looked out on the pond again. Yes, I was excited to see it—I was excited to have made it to Walden—but naturally it was not any more or less beautiful than a lot of ponds I have seen. It looked like a large kettle pond surrounded by trees. The water seemed pleasant. I was feeling a little guilty about not being particularly moved in one way or another about the pond's beauty, but at a certain point, it occurred to me that thinking about it as a no-big-deal pond pleased me too. Upon even further reflection, I came to see that "no big deal" is the point of Thoreau for me. Put another way, the smaller the deal the bigger the deal. Thoreau is my patron saint of opposites. Emerson thought he was mostly being contrary; I think he was trying things on inside out to see not just how they fit, but how they are made.

THROUGH THE YEARS, Thoreau has meant a lot of different things to different people who have gone to Walden Pond, beginning for the most part with women. Early on, Thoreau societies were composed mostly of women; nature writing, as a genre, was mostly the province of women in the years after the Civil War. Men visited, though. John Muir came to Walden, in 1883, to lay flowers on the graves of Emerson and Thoreau, and then to walk to Walden. It was a promotional tour of sorts, as he was being led around by his editor. "No wonder Thoreau lived here two years," Muir wrote later in a letter. "I could have enjoyed living here two hundred or two thousand. It is only about one and a half or two miles from Concord, a mere saunter, and how people should regard Thoreau as a hermit on account of his little delightful stay here I cannot guess."

In the twenties, Thoreauvians began to break into two (sometimes overlapping) camps, seeing Thoreau as the nature writer or as social critic. A popular pamphlet series at the time described Thoreau as "The Man Who Escaped from the Herd," and in some less conservative magazines he was championed as one of the first "native" writers to criticize U.S. imperialism. In 1926, Lewis Mumford, in *The Golden Day:*

A Story of American Experience and Culture, presented Thoreau as an undervalued critic of American materialism. For its part, materialism thrived—in 1925, Calvin Coolidge said, "The business of America is business"—until the Great Depression, which probably increased *Walden*'s audience. "Thoreau is the only author I know of that I can read without a nickel in my pocket and not feel insulted," Walter Harding quoted a down-on-his-luck friend as saying.

During World War II, Thoreau disappeared to some extent; it was not a good time for dissent, or so it seemed. As the war began, the *New York Times* wrote, "Today the nation demands civil obedience for national defense."* (Meanwhile, European resistance groups were reading Thoreau for inspiration during Nazi occupation.) By 1945, the centennial of Thoreau's trip to the pond, editorial writers cited him as an indirect cause of World War II: "Is there anything in his philosophy that would have corrected the isolationism of the Thirties, when, like Thoreau, we withdrew from the society of nations and tried to preserve our freedoms by living to ourselves? At a terrible cost we have learned, and shall try to remember always, that we cannot be free by trying to live alone."**

Like a shrub that grows back when cut to the ground, the conception of Thoreau became more substantial in the

* Arthur Schlesinger, in his book *The Age of Jackson,* warned about pacifists who were not as principled as Thoreau: "Little men covering cowardice with a veil of self-righteousness, lay claim to the exemptions of a Thoreau with the most intolerable pretense."

** F. O. Matthiessen was the first to describe Walden's metaphorical structure, in 1941, and looked at Thoreau's process of creation, but found that his self-sufficiency and independence was "without any validity to modern civilization." Walter Harding wrote: "There is no better antidote for *Mein Kampf* than *Walden*." (Harding was a conscientious objector.)

fifties, stronger, thicker, as the masses became more massive. "Henry Thoreau in Our Time," an influential essay by Stanley Edgar Hyman, described Walden as a "vast rebirth ritual" that encouraged collective action on the part of the individual. Opponents of the House Un-American Activities Committee and McCarthyism referenced Thoreau in his Concord jail in a full-page newspaper ad in 1951. After Rosa Parks refused to give up her seat on a bus to a white man, Martin Luther King, Jr., a Thoreau disciple via his reading of Mahatma Gandhi, used a protest device referred to as nonviolent resistance. It was a variation on Thoreau. "I had become convinced that what we were preparing to do in Montgomery was related to what Thoreau had expressed," King wrote in *Stride Toward Freedom*. "We were simply saying to the white community, 'We can no longer lend our cooperation to an evil system.' "*

By the sixties, Thoreau was recovered by the earliest members of the antimaterialist movements. "The failure of material success to produce happiness has brought a new appraisal of values that is turning more people to those of Thoreau," a political scientist wrote. And Thoreauvian civil disobedience had reached into different causes, such as nuclear disarmament. In the midst of nuclear weapons protests, the rowboat used to picket nuclear submarines in Groton, Connecticut, was named *Henry David Thoreau*. The idea of Thoreau as the man who went to a pond was in large part

* You can work any writing to your advantage, but surely even managerial professionals must have felt as if they were reaching when Thoreau was pitched as a manager in "Management Man" (*Nation's Business,* February 1949), and when he was paraphrased in an advertisement for Revlon cosmetics in *Business Week* on August 12, 1950: "Most men lead lives of dullness, quiet desperation, and I think cosmetics are a wonderful escape from it."

replaced by the man who spent the night in jail, and people argued over whether or not he might have abhorred or supported violence. The Thoreau Society supported César Chávez's United Farm Workers' nonviolent protests in 1969. In 1970, a society member protested at the annual meeting, saying the group was becoming "a House-and-Garden Club" instead of dealing with "the problems of modern America." He wanted protests against the Vietnam War, condemnation of racist politicians, and a demonstration "*en masse* at the business of some major polluter."

I DON'T KNOW HOW OTHER PEOPLE DO IT, but I wanted to take the trail along the edge of the pond to get to Thoreau's house site, as opposed to the more direct-seeming one, through the woods, and I started out that way, but somehow I got screwed up and ended up backtracking, and then I was back almost to where I started. At that point a couple asked me for directions. "Sorry, it's my first time," I said. We all nodded and continued in our different directions.

As I walked the trail, I knew, of course, that wildness, in the popular sense, was not what I was observing. I understood this at a cursory glance, but additionally, I had spoken one day with a biologist, Richard Primack, who had been working in the area. His aim was to compare the dates and times of, for instance, plant flowering in the past with the dates and times of plant flowerings now—also in response to global warming. He came upon Thoreau's charts, graphs, and notes, the pages and pages of pure measurements and various ecological information that Thoreau had compiled in his journal. Thoreau's zeal for rote recording of plant flowerings or water levels, of the date and time of the appearance of buds and leaves and sprigs, is like gold to a modern environmental biologist: it lives on, still blooming (and is proof, in the short term, of the drastic effects of global warming). "How sweet is the perception of a new natural fact!" Thoreau wrote in his journal in 1852.

Primack told me that when he is taking people on hikes to look at the plant species that interest him, for instance, he is less likely to take people toward Thoreau's house site or the trails themselves. "Those areas are actually not so interesting, biologically," he said. He is more likely to take them near the parking lots or, if it's just him, toward that railroad cut that Thoreau so loved. "There's a greater diversity of species when you have the intersection of the human and the natural world.

"He was always looking for the wild in a human-dominated landscape," Primack added.

Of course, the degree of human dominance of the landscape in Thoreau's time was different. The Thoreauvian search for what's wild within what is seemingly not is now—with acid rain, global warming, pollutants in water supplies and food chains everywhere—more ubiquitous, as much of an option at the parking lot in Walden as in a national park.

In a way, Walden Pond seems small to me compared to Thoreau's view of nature, which is even larger than I might once have thought. The problem—if I have not said this too many times already—with the Thoreau so many people know is that he perpetuates a separateness between man and nature. We see the nature of Walden Pond as separate from the nature of the railroad tracks. We see the nature at the beach on the pond, which we try to keep litter free, as separate from nature in our driveways, where our car has a leak and the oil seeps out and down into the street and away to who knows where—maybe the rivers, the watershed, a landscape larger than any subdivision or municipal jurisdiction, an area that ignores national boundaries. We see our individual actions as separate from the actions of our community, economic or recreation or otherwise, when we are all creatures in the same landscape, a herd, a mass of men and women. With a Thoreau who is separate from us, then we don't see our actions, the *how* we live, as relating to Thoreau's nature, which is in town, right where we live.

I was walking through Walden Woods—*the* Walden Woods! I was exhilarated. But at the same time I knew that *Walden* reminded me that there is no place any more special than another. The understandable human tendency in nature writing to celebrate the extraordinary in the natural environment makes other places seem less than extraordinary, or bad, or ratty. Surely, we have our evolutionary reasons: great views may be linked somehow, deep within our memories and genes and bones, to defense of home and community. Proximity to lovely-looking water bodies bodes well for survival, given the likelihood of fish, wildlife, and water. But exceptionalism ends up being un-Thoreauvian, as well as unfair and unjust, as applied to other species, or places, or even neighborhoods. Exceptionalism leads to trash incinerators being sited in low-income neighborhoods and eventually to the loss of the mundane that makes the extraordinary possible.*

* There are lots of good things about living in the country, vis-à-vis romantic nature and its appreciation. If there is a good thing about living in a city from the point of view of the conception of nature in our day and age, it may have to do with the reverse commute theory, for we are well placed to notice the things that others might take for granted. A hawk that lives on an apartment building

I live in the city and was visiting the woods, and came to an intersection and saw a sign for Thoreau's house site—I was close—and I was getting the feeling more and more that in the city I might find nature in a lot more places than I might have looked for it before, that the *wild*ness was more important than wilderness, that wildness was everywhere, if I looked for it, the search being part of what makes wild wild. "In wildness is the preservation of man," Thoreau wrote.

To turn to the trend of Thoreau's day that even Thoreau could not escape—etymology—*nature,* as Thoreau would have investigated, comes from the Latin *natura,* meaning "birth," and *nasci*, "to be born." We are born into nature, into all, into the universe. The difference is in what we see, or our *environment,* a word that comes from the Old French and means "all around," as in everything, everywhere, as in the nature at Walden and the nature everywhere else.

I WAS WALKING FAST, so I stopped once in a while, not because I was tired or out of breath, or even really wanted to at all, but because I felt I should occasionally, the way you might feel as if you ought to take your hat off in a church, or the way you might hold a door for someone, just out of respect. I came into a little hollow that I liked for its apparent seclusion, even though I imagined that on a weekend in summer the foot traffic would be competitive with that of a shopping mall.

Then, I came upon three young women and a man, all of whom were in their twenties. The man was leaning over,

---

is no more unusual than a hawk that lives on a cliff; it's the same hawk. It's not an extraordinary event for the hawk at all; it's a pragmatic solution. It can be, though, an extraordinary opportunity for the city-dwelling human.

with a small frog in his hand: it was just graduating from tadpole. He was placing it into the pond from the cove side of the log bridge that we had all walked across. He had apparently gotten the tadpole into some kind of danger, before I came upon the scene, or so he felt. "I don't normally touch the wildlife," he told his companions, "but I feel like this was my fault to begin with." All four people were wet, as if they had just been swimming. I said hello, but we didn't talk much. I am the kind of person who has to try to not be excessively sociable.

WHAT WOULD THOREAU HAVE DONE WITH THE frog? What would Thoreau do, moreover, if he were visiting Walden Pond today? Would he have walked or driven or not gone at all? And would he have insisted on an all-organic lunch? Or gone with what they were serving at the local sandwich shop, as I did? Would he have taken a train or public bus? Would he have come in a town car while on an author tour? On the one hand, these are those unanswerable hypotheticals—along the lines of asking whether George Washington would cut taxes on corporations or if Jesus would give up his seat on a crowded flight if he was in an exit row. On the other hand, Thoreau has anticipated the questions, in his way. His Walden "experiment" was like a science experiment, set up specifically to pose a question, designed to move on from hypothesis, and, as far as questioning goes, to inspire even more. The question he was ultimately working with supersedes questions about whether he abhorred technology (which he didn't) or society (ditto) or whether he would have used a solar-powered charger for his cell phone—that question being: How should you live (emphasis on *you*)?

But you can ask what he would do, if you feel like it.

Remember that he was living in an age of reform, at the dawn of magazines whose writers purported to know what you ought to do, what you ought to buy, exactly how you might purchase the things that you would need to purchase in order to simplify. Consider that vegetarianism, or eating right, and supporting local farms, trends now, were trends then. Consider that there were best-selling authors and well-read journalists who proposed that we reform how we eat and grow food and where we buy it and that Thoreau partook in a lot of those trends, but at the same time, had a laugh at that, working in his own experimental garden behind the shack that was about all he could afford, anyway, as his contemporaries went off to their second homes where they hired people (like Thoreau) to grow their food, per the trend, among an expanding list of other things that people hired other people to do.

Thoreau doesn't offer answers. His is the analysis that leads to the questions. For application purposes, you can apply Thoreau to any question, not to find the answer, but to imagine how he might pose it anew. When you ask what car to drive, imagine Thoreau asking where you are going, or if the car is driving you. You can watch the word *green* as it enters the language through the Saxon (the *grenost* leaf), is used by Shakespeare ("How lush and lusty the grasse looks? How greene?"), and then transforms from the green of the youthful employee to the green of a green thumb to the green, in the early 1970s, that was associated with the German political groups opposed, first, to nuclear power and then to the introduction of, for example, genetically engineered crops, as well as illegal immigrants. Thoreau's not necessarily part of a green movement; he's the guy who wonders where the movement's moving to and asks what green means, anyway? He's not reactionary, or he was trying not to be. He's prophetic. That doesn't mean that Thoreau was trying to predict the future—to determine the extent of global warming—or say what a president or congressman or neighbor should or shouldn't do. He was just trying to say something that's true.

Would he blog? Would he buy fish only from sustainable industries when ordering seafood at a restaurant? Would he allow his children—if he had any, or if he eventually adopted—to eat nonorganic baby food? Would he use Google or Yahoo! or another Web browser, since Google depends on hydroelectric power that affects salmon runs in the Pacific Northwest, and would he use a BlackBerry in a restaurant or wait until he got outside or back to his hotel room? Most crucially, perhaps, what kind of car would Thoreau drive?

"But lo! men have become the tools of their tools," he writes in *Walden*. "The man who independently plucked the fruits when he was hungry is become a farmer; and he who stood under a tree for shelter, a housekeeper. We now no longer camp as for a night, but have settled down on earth and forgotten heaven."

Would he advocate flexible work time, a third political party, big-box stores in shopping malls if they have good health insurance for their employees and use recyclable packaging?

Walden again: "There are such words as joy and sorrow, but they are only the burden of a psalm, sung with a nasal twang, while we believe in the ordinary and mean. We think that we can change our clothes only."

A THOREAU-RELATED QUESTION that I like to think about when walking and thinking about Thoreau—as I was, obviously, as I neared the final minutes of my journey to the site of his house on Walden Pond—is, If I could walk on any of the walks that Thoreau took in his life, where would I walk with Thoreau? I love the woods, and I would not mind having a small house on a pond, but if I could choose any walk with Thoreau, I would choose a walk that is less reported. There are books of Thoreau's treks into Maine and New Hampshire's White Mountains, and Thoreau's walk to Wachusett is one of the most popular hikes in Massachusetts today, but my favorites are the ones he took with Walt Whitman in New York. When Thoreau knew him, Whitman walked from Plymouth Street, in Brooklyn, where he lived with his parents, to the landing for the ferry to Manhattan, the landing being very close to where I live today. On more than one occasion, Thoreau apparently joined him. When I think of them together, the ultimate city poet and the ultimate nature writer, the divide between city and country, between nature and civilization, melts away like a polar ice cap.

Together, they were the opposite sides of the same national coin, one man looking for the future of America in crowds on the street, the other looking for it in the ice crystals forming on the side of a railroad track on the edge of a hacked-up woodlot. Whitman sought the great equanimity of mankind in his leveling lists; Thoreau saw equanimity in the course from life to death, the universal return to nature, the descent and then ascension from the great mire, the muck and dreck from which life miraculously sprang. They both tried to play in the key of joy, both being extra vagrants, both chanticleers, both preaching the same or similar sermons but from different pulpits, with dramatically different styles, different wits. Whitman sang the body electric looking out on the crowd of new faces, loving his body, or trying to. Thoreau was the easily startled rooster, waking up the neighbors, his body charged but shocking still even to him. Whitman speaks to us all at once, a group hug; Thoreau speaks to us one at a time. When I imagine them walking together, I see them stepping out of Whitman's house, past the lilac bushes that Whitman's parents grew near the front door.

Walking was the specialty of both. Whitman would write and then lunch and rest and walk, walking into the nooks and crannies of the city, or ride the ferries, or the carriage bus up Broadway, where he eventually met his lover. Whitman rode the carriage up and down Broadway just for the view. It was the city equivalent of touring the farm, of visiting the county fair. The man on the street was to Whitman what the farmer was to Thoreau, and when he was alone, Whitman liked to go to the shore and read Homer. They both talked about living outside the commercial world. But they straddled both, each man's work a kind of Talmudic response to antebellum society.

Whitman thought Thoreau too stiff in a way; in another way, he liked Thoreau's forthrightness—when he showed up unannounced at the Whitman family household, Thoreau just walked back to the kitchen to meet Whitman's mother, pulling food from the oven, not embarrassed at all. "You would have to like him," the great democratic man reminisced in his old age. "He was always doing things of the plain sort—without fuss. I liked all that about him."

The reports indicate that the first time they met, they stared at each other; an awkward pause ensued. Thoreau was visiting New York, and he stopped by Whitman's to meet the man who Emerson had famously praised—"I greet you at the beginning of a great career." Without asking, Whitman had used the line from Emerson on the spine of the next edition of *Leaves of Grass,* and had the letter printed in its entirety in the *New York Tribune.* Whitman impressed Emerson more than Thoreau or Thoreau's books ever had. Thoreau had come to Brooklyn with Bronson Alcott to hear Henry Ward Beecher preach. (He was not overly impressed. "If Henry Ward Beecher knows so much more about God than another, I would thank him to publish it in Silliman's Journal, with as few flourishes as possible," Thoreau remarked.) The next day they called on Whitman. With them was Sarah Tyndale, an abolitionist and Fourierist from Philadelphia, who would subsequently befriend Whitman.

It was a cautious meeting, in Whitman's garret, the walls decorated with somewhat shocking drawings: a satyr, Bacchus, Hercules, images of gods that Whitman seemed to channel as a poet. The bed was unmade, the impressions of Whitman's and his brother's bodies still apparent, the bedpan was in the open. Thoreau was reserved and contrary, as usual, though contrary in a way that was mostly complimentary to Whitman. Thoreau called Whitman's critics "reprobates." Whitman was suddenly taken aback and said he thought "reprobates" was severe.

"Do you regard that as a severe word?" Thoreau said, as remembered years later by Whitman. "Reprobates? What they really deserve is something infinitely stronger, more caustic. I thought I was letting them off easy."

Whitman mostly talked about Whitman; Thoreau mostly listened but, in the mind of the bystanders, didn't so much talk as stalk the corners of the conversation. Each was wary, waiting to pounce. Later, when Whitman would visit Emerson in Boston, Emerson would walk him around town, attempting to convince him to tone down his enthusiasm for the body in his poems, even though Whitman was, in a way, doing what Emerson's writing had prescribed: exhaling poetry after inhaling the world's soul.* But when Thoreau got home, he wrote a letter to his pen pal, H. G. O. Blake, and called Whitman "the most interesting fact to me at present." Thoreau carried around *Leaves of Grass* "like a red flag," Emerson noted. Thoreau especially loved "Crossing Brooklyn Ferry"; indeed, it achieved in a few pages what Thoreau had tried to do in his book *A Week on the Concord and Merrimack Rivers*. Thoreau saw it as better than any Beecher sermon, and though he was nervous about the sensuality of *Leaves of Grass,* he praised it in the end. "We ought to rejoice greatly in him," Thoreau said.

Emerson could tell Whitman had had an effect on him, and he mentioned it in his eulogy, though Sophia Thoreau made him strike the reference—more fear of eccentricity! When Thoreau got home from the visit, he debated the merits of society versus solitude with himself and friends, and still preferred solitude, or his friend-laced version of it, but when a fan wrote to ask Thoreau for a reading list, he quoted Confucius: "Conduct yourselves suitably toward the persons of your family, then you will be able to instruct and direct a nation of men." He was seeing solitude as a means to a more social end.

In 1888, after Thoreau had died, and Whitman had had

* The Thoreau scholar Lewis Hyde has noted that when Whitman tried to raise money later for Civil War hospitals, a friend in Boston told him that the Transcendentalists were prejudiced against him: "It is believed you are not ashamed of your reproductive organs and, somehow, it would seem to be the result of their logic—that eunuchs only are fit for nurses." In *The Gift: Creativity and the Artist in the Modern World*, Hyde writes, "We might say that Whitman was Emerson's enthusiast, Emerson with a body."

a stroke and moved to Camden, New Jersey, Whitman recalled Thoreau as less precious than Emerson and as more of a force: "He looms up bigger and bigger," Whitman said. "His dying does not seem to have hurt him a bit. Every year has added to his fame." I like the fact that Whitman felt associated with him *because* of his dissent: "One thing about Thoreau keeps him very near to me: I refer to his lawlessness—his dissent—his going his own absolute road let hell blaze all it chooses."

At last, about an hour after leaving the replica of the Thoreau house at the Concord Museum, I arrived at the original site of the original Thoreau house. It was great: quiet, bucolic, a view of the pond through the trees, Thoreauvian-free—*just me alone, to think!* The pile of rocks was there, a cairn said to have been begun by, in one story, Lidian Emerson, but by other people in other stories. All trace of the house is gone, its ruins discovered and excavated in 1945. A plaque marks the spot:

Site of Thoreau's Cabin
Discovered Nov. 11, 1945
by
Roland Wells Robbins

I stood as if I were in the cabin, as if I were at the front door and looking out over at the pond. I tried to imagine it without so many trees, to have in my mind's eye what it looked like around the time Thoreau was there. I tried to imagine the Irish guys living in shacks and working for the railroad, just before he arrived. I tried to imagine the manual laborers working for the ice company, hacking away at the ice, falling in the water, shouting, freezing, warming by Thoreau's fire. I could see how excited Thoreau might be when Therien the woodchopper stopped by, when his friend Channing spent the evening, when he put that extra chair out in front of his house and, seeing it, company came. I heard people laughing. I imagined a watermelon party and an abolitionists' meeting. I pictured him locking his desk as he set out for town, worried about his manuscripts, his journal. I tried to imagine Emerson walking up, saying "*Henry?*" His other friends, playing on his given name, saying, kiddingly, "*Dave?*"

I could imagine how wonderful it would be to work alone, early in the morning or late at night, with little in the way of social distractions. I could readily imagine falling in love with the pond—its water, the surface in the morning, a fish hawk perching on a dead limb, watching quietly. Once, my family and I rented a house just up the river from New York City, in some woods, and there was a pond that we walked to every day to look at, to think, to live. Nature!

BEFORE I LEAVE WALDEN, or recount my leaving, and before I close this book, I'd just like to mention that on a similarly beautiful spring day, a Sunday morning, I went out on a walk with my wife and daughter to a spot that I feel certain Thoreau and Whitman might have passed on one of their walks together. (My son was off practicing with his band.) It's in Brooklyn, a little more than a mile from my house, and closer to downtown, in a neighborhood called Vinegar Hill. Vinegar Hill was named for the 1798 battle for Irish independence, the realtor hoping to attract more Irish immigrants in addition to those who were already there. (The name Vinegar Hill comes from an English transliteration of an Irish phrase that translates roughly as *hill of the woods with berries.*) In Thoreau and Whitman's time, it was a neighborhood filled with immigrants and workers as well as freed slaves, the kind of neighborhood Whitman loved to wander through. It was at the edge of a large residential neighborhood that stretched from the Brooklyn Bridge to the Brooklyn Navy Yard, a neighborhood that now is mostly not there, that has been paved over with roads and ramps and public housing.

On that spring Sunday, we walked across Sand Street, once known as Hell's Half Acre because of the brothels and bars, now a glorified highway on-ramp. It's what the newspaper real estate pages call a neighborhood in transition, which means that new money is pushing the old residents out, a natural succession, of sorts, given the economic climate. We passed some old public housing, and we passed some new luxury condominiums, mostly luxury-tenant-less. We passed artists' studios and galleries, and a few old shops manufacturing things that would soon be manufactured no more. We passed a Buddhist temple, built on an old car-repair lot. We made our way to Plymouth Street and Hudson, where there exists a tiny remnant of nineteenth-century houses and buildings, located smack up against the Con Edison electrical plant between Vinegar Hill and the East River.

This is what I wanted to show my family. I love this little row of houses. I love it because it feels like a secret, a time capsule into Whitman's Brooklyn, the Brooklyn of slow walks and hard work, and because of its powerful juxtaposition to the security-camera-covered, high-voltage Con Edison plant spewing microscopic particulate matter into the air. Specifically, I love the little buildings and their lilacs. These are my favorite lilac bushes in all of Brooklyn, if I may say so. These lilacs are off the beaten path, for sure, but I travel out of my way to see them from time to time, when I know lilacs are in bloom. They feel like the center of an old village, a village that no longer exists. They are not exceptional lilacs; they are excitingly *un*exceptional. From all I have read, from every historian and history buff and interested personage that I have talked to, I feel certain that Whitman and Thoreau would have stood on this corner, if only for a second, looking down toward the East River. The lilacs may be new, or they may contain the DNA of Whitman-era lilacs. I don't want to know. I just want to think of all that comes together for me when I stand there smelling those lilacs, lilacs being to Thoreau a symbol of man that lingers in the land. From his journal:

> Still grows the vivacious lilac for a generation after the last vestige else is gone, unfolding still its early sweet-scented blossoms in the spring, to be plucked only by the musing traveler; planted, tended, weeded, watered by children's hands in front-yard plot,—now by wall-side in retired pasture, or giving place to a new rising forest. That last of that stirp, sole survivor of that family. Little did the dark children think that that weak slip with its two eyes which they watered would root itself so, and outlive them, and house in the rear that shaded it, and grown man's garden and field, and tell their story to the retired wanderer a half-century after they were no more,—blossoming as fair, smelling as sweet, as in that first spring. It's still cheerful, tender, civil lilac colors.

I showed them to my wife and daughter, and then we all stopped and smelled the lilacs.

BACK AT THE POND, A LITTLE TIME PASSING, no one came as I stood there at the site of Thoreau's experiment. I was still alone. It was all very tidy, no litter, very respectfully visited, it seemed to me, though again, it was not a high-traffic day. I took a photo with the camera in my cell phone and pushed *send* so that it went off to my wife and then to the computer of my friend who doesn't have a cell phone.* I took no calls during my time there, though none came through. Instead, I marveled at my good fortune, having a job that allowed me to just stand there, a playful work. Also, I read the large plaque:

> I WENT TO THE WOODS BECAUSE I WISHED
> TO LIVE DELIBERATELY, TO FRONT ONLY THE
> ESSENTIAL FACTS OF LIFE, AND SEE IF I COULD
> NOT LEARN WHAT IT HAD TO TEACH, AND NOT, WHEN I
> CAME TO DIE, DISCOVER THAT I HAD
> NOT LIVED. I DID NOT WISH TO LIVE WHAT WAS NOT
> LIFE, LIVING IS SO DEAR; NOR DID I WISH TO PRACTICE
> RESIGNATION, UNLESS IT
> WAS QUITE NECESSARY.

I was determined to savor the moment, though I was getting antsy, as is my wont. I was torn. I wanted to stay, but I wanted to start back, to relax at the old hotel, have a beer, sit on the porch, watch the people go by, deliberately relax with essentials in a setting that was less deliberate, less contemplated. What I mostly wanted to do was transport in

*He finally bought one the other day, but when I called him on it, he still hadn't figured out how to use it.

some family and friends and do the same with them at the site of Thoreau's house—it is a great picnic site, a crowd-pleasing spot, as Thoreau apparently knew well.

The last words that Thoreau wrote in his journal were about his beloved railroad cut; he managed to get himself out of bed one last time, trudge over to the tracks in the woods after a rainstorm, and describe the impression made by the rain in the sand. Then, after twenty-five years of reporting to it, he wrote his journal's last line: "All this is perfectly distinct to an observant eye, and yet could easily pass unnoticed by most." Thus, as I started back, I looked around one last time, trying to figure out what I wasn't seeing—I wanted to be the Dr. Livingstone of that midweek moment at Walden Pond.

That's when I looked down and saw it—or discovered it, I should say. A bar of music, all scratched out in the sandy ground as if the dry dirt were music paper. The breeze had made it nearly illegible, but I could tell that in the tune, the just-about six notes were ascending. It was a happy tune, in other words, joy. What could be better than to end a book about Thoreau with a little snatch of a tune? If you are me, thinking about Thoreau the way I do, then nothing. That's what's missing in the visit to Walden Pond, the soundtrack—the sound of Thoreau playing the flute, not long, slow, mournful tunes, but the little trills and happiness that you have to play if you have a flute, the jigs and reels that everyone in Concord and even America knew at the time: You feel the two-way street of music, the sounds bouncing off the woods and nestling back in you, pure, like spring water, solitary *and* social. At the pond you miss the sound of laughter and applause, of hands being clapped on the singer's back, after Thoreau sings his favorite song, "Tom

Bowling," a sad song about a drowned sailor, a good guy who had gone to a better place. Thoreau must have thought about his brother when he sang it, which he did often. It was the song he sang for friends. People gave him the sheet music, then they requested he sing it whenever he came to visit or stay. You can think that sad songs make people sad, or you can recognize that when a guy sings a sad song at a party, as Thoreau often did, everyone leaves a little happier, maybe a little better off. Music is joy, in a minor key or a major one, and surely joy is the condition of life.

Acknowledgments

GILLIAN BLAKE; Molly Lindley; Cecilia Hunt; Eric Simonoff; Gerald Marzorati; Vera Titunik; Jin Auh; Tracy Bohan; Andrew Wylie; Sara Mercurio; Benjamin Adams; Karen Rinaldi; Anna Wintour; Sally Singer; Laurie Jones; Alexandra Mack; Florence Kane and Michael Mraz; Ned Martel; Marshall Mockler; Eamon O'Leary; Dana Lynn; the Brass Monkey; Susan McKeown; Townsend Fuerst; Eric Etheridge; Jay Rabinowitz; Daphne Klein; Chris Mellon; Manny Howard; Gabrielle Howard and Marty Skoble; Saint Ann's School; Two Fiddles; Two for the Pot; Skip McPherson; Chris Knutsen; David Foster and the Harvard Forest; Charlie Butler; Anthony Andreassi; the 7 @ 7 Club; Brooklyn Public Library; Pete's Waterfront Alehouse; The DesignWorks Group; James Leinfelder; Dennis Lichtman; Brenda Marsh and Jonathan Weiss; Paula Greif; Foghorn Stringband; Peter Watrous; Scott Menchin; Book Court; D'Amico's; Rich Pliskin; Andrew Wagner; Dan Barry; Geoff Brewer; the Oratory at Saint Boniface; Victor Marchioro; Lloyd Miller; Colm Tóibín;

Maria Falgoust; the Quinns; the Sugar Pond Community; Kassie Schwan and Brian Rose; the Kaminsky family; the Diehl family; the Sullivan families; Matthew "Matt" Sharpe; Jill Desimini and Dan Bauer; Linda and Donald Desimini; Mary Elizabeth and Robert Sullivan; Louise Sullivan and Sam Sullivan.

Notes

BOOKS I CHIEFLY RELIED ON, for the arc of Thoreau's life and for insights into his thinking, were *The Days of Henry Thoreau,* by Walter Harding, and *Henry Thoreau: A Life of the Mind,* by Robert D. Richardson. The Harding book is old, first published in 1965, but it still holds up; it is dedicated to Martin Luther King, Jr., and Edwin Way Teale, an activist and a naturalist, both Thoreauvians. Richardson's book is like a live video feed of Thoreau's thinking, the cultural and intellectual context an education in itself, the typesetting and woodcuts (by Barry Moser) unusually beautiful in a book so learned. To keep straight on the day-to-day activity of Henry Thoreau, I spent many satisfying days in *The Thoreau Log: A Documentary Life of Henry David Thoreau, 1817–1862,* by Raymond R. Borst. *Prophet in the Marketplace:*

Thoreau's Development as a Professional Writer, by Steven Fink, especially helped me with Thoreau's early career, as well as detailed Thoreau's efforts to strike that ever-elusive balance between writing what you want to write as a writer and writing what will sell. Also helpful in this area, as well as for understanding Thoreau's temperament as a freelancer, was *The Book of Concord: Thoreau's Life as a Writer,* by William Howarth, who writes: "Thoreau's genius was for nonfiction, the sort of creative journalism that has flourished in America, producing writers as diverse as Agee, Didion, McPhee and Mailer. Thoreau launched this tradition; his career stands as a reminder that no fact is trivial, if seen in the proper light—and with an observant eye." I referred to my two-volume Dover edition of Thoreau's journals—fourteen volumes, bound as two giant books—and, later, the excerpts of his journal edited and annotated by Jeffrey S. Cramer and collected in a beautiful edition by Yale University Press, *I to Myself.* I also checked in online with the journals as they are being retranscribed by Princeton University Press, an amazing projected sixteen-volume project that was begun in 1981 and is not yet done, because of Thoreau's handwriting, among other things. Elizabeth Witherell runs the Thoreau Textual Center at the University of California, Santa Barbara, and you can see pages of the journal online: http://www.library.ucsb.edu/thoreau/.

Other books on Thoreau in general that I referred to throughout include: *Thoreau: A Collection of Critical Essays,* edited by Sherman Paul; *The Magic Circle of Walden,* by Charles Roberts Anderson; *Henry Thoreau, as Remembered by a Young Friend,* by Edward Emerson, who was the son of Ralph Waldo Emerson; and *The Environmental Imagination: Thoreau, Nature Writing, and the Formation of American Cul-*

ture, by Lawrence Buell. I read a lot about Emerson, because to understand Thoreau you have to have a pretty good handle on Emerson. I referred to *The Life of Ralph Waldo Emerson,* by Ralph L. Rusk, and the article "Big Dead White Male," by John Updike, in the August 8, 2004, issue of *The New Yorker.* Mostly, I studied *Emerson,* also by Lawrence Buell, especially in regard to Thoreau and Emerson's synergistic (and nonsynergistic) relationship. *American Transcendentalism: A History,* by Philip Gura, shows the resilience of Transcendentalism hidden even in American culture today, and it is by an author I enjoy for, among other things, his wide range: see his books on the theologian Jonathan Edwards and the history of the banjo, as well as his short history on the New Lost City Ramblers, an old-time string band begun in the late 1950s by Mike Seeger, who was, incidentally, a conscientious objector—in Thoreau speak, a resistor—during the Korean War and fulfilled his alternative national service as a dishwasher in a Baltimore tuberculosis hospital.

In the introduction, "the prophetic voice" is a phrase that comes from Lewis Hyde, whose essay "Prophetic Excursions" introduces the North Point Press edition of *The Essays of Henry David Thoreau.* The comment from Thomas Wentworth Higginson comes from *Towards the Making of Thoreau's Modern Reputation: Selected Correspondence of S. A. Jones, A. W. Hosmer, H. S. Salt, H. G. O. Blake and D. Ricketson,* edited by Fritz Oehlschlaeger and George Hendrick. J. B. Jackson's essay on Thoreau is "Jefferson, Thoreau and After," included in the collection *Landscape in Sight.*

For information about Concord at the time of Thoreau's upbringing, I relied heavily on the writing of Robert Gross, in addition to that of Harding and Richardson. When you write a book and research a topic, you often run into a few writers whom you are pleased to meet, whose writing you cannot believe you had not previously read. So it was with me and Gross, whom I don't know except on the page. I relied, for instance, on "Commemorating Concord: How a New England Town Invented Itself," published in *Common-place,* an online journal (vol. 4, no. 1), in October 2003. "That Terrible Thoreau?: Concord and Its Hermit" is an essay by Gross in *A Historical Guide to Henry David Thoreau,* edited by William E. Cain. I recently read an only somewhat Thoreau-related book by Gross, *The Minutemen and Their World,* which included a line in the acknowledgments that reminded me of *Walden,* despite Thoreau having been a bachelor: "Matthew Gross did nothing to help publish this book; indeed he delayed the completion of this manuscript. For that, I am most grateful. He took his father away from his work, gave him much pleasure and joy, and led him to realize what was truly important and what was not."

The quotation from Horace is from a translation by Derek Mahon and featured in *Derek Mahon: Selected Poems,* which is also home to "A Disused Shed in County Wexford," a beautiful and, to my mind, very Thoreauvian poem. The quote about poverty making food taste better comes from *The Unitarian Conscience: Harvard Moral Philosophy, 1805–1861,* by Daniel Walker Howe. The quote from a letter to Helen, with the pun on familiarity, comes from *Concord* magazine, in an article written by Richard Smith, who sometimes portrays Thoreau in the Concord area. The quote from Bob Dylan comes via Todd Haynes, who used it in his film, *I'm Not There,* having excerpted it from a *Playboy* interview conducted by Nat Hentoff. The translation of the *Georgics* I referred to is a beautiful one by Janet Lembke.

The information about forests in Thoreau's time comes

from the work of David R. Foster, including his books *New England Forests Through Time: Insights from the Harvard Forest Dioramas* and *Thoreau's Country: Journey Through a Transformed Landscape*. The effects of railroads on America came from sources such as *Zoomscape: Architecture in Motion and Media,* by Mitchell Schwarzer. Technological changes emphasized in a May 8, 2007, *Time* article by Kurt Anderson summarize the technological backdrop of his novel *Heyday,* which presents an antebellum America that in temperament feels a lot like America now. Information on Orestes Brownson came from *Orestes A. Brownson: American Religious Weathervane,* by Patrick W. Carey, and many of the quotes from the Transcendentalist writings can be found in *The American Transcendentalists: Essential Writings,* edited by Lawrence Buell. Buell discusses the way in which the Transcendentalists made a radical call for nondogmatic experimentation. Information on America's interest in wordplay and etymology comes from *Transcendental Wordplay: America's Romantic Punsters and the Search for the Language of Nature,* by Michael West, which opens with this caution: "Tons of puns—that's what this book contains. Fair Warning!" I did not get into Thoreau's interest in scatological puns, but it is discussed there if you'd like to delve into it on your own.

In describing the economic scene and how it related to antebellum American culture, I referred to *The Market Revolution: Jacksonian America, 1815–1846*, by Charles Sellers, which was reportedly too radical for the Oxford University Press's History of the United States series, but not too radical for me. Among the articles I referred to about labor and working conditions were: "Ante-Bellum Labor: Violence, Strike, and Communal Arbitration," by David Grimsted, published in the *Journal of Social History,* Autumn 1985; "Healthful Employment: Hawthorne, Thoreau, and Middle-Class Fitness," by Michael Newbury, in *American Quarterly,* December 1995; "Living on the Boott: Health and Well-being in a Boardinghouse Population," by S. A. Mrozowski, E. L. Bell, M. C. Beaudry, D. B. Landon, G. K. Kelso, in *World Archaeology,* October 1989; and "Religion and the New England Mill Girl: A New Perspective on an Old Theme," by Jama Lazerow, in the *New England Quarterly,* September 1987. Eight hundred women marched the streets of Lowell in 1834, upset about factory wages, singing this song, which itself shows: (1) that once people considered fair wages to be tied up with the idea of patriotism and not just the bottom line, and (2) that once protest was considered patriotic:

Let oppression shrug her shoulders,
And haughty tyrant frown
And little upstart ignorance
In mockery look down.
Yet I value not the feeble threats
of Tories in disguise,
While the flag o independence
O'er our noble nation flies.

Concerning the relationship between Emerson and Thoreau, and relationships between nineteenth-century men in general, I read "Romantic Friendship: Male Intimacy and Middle-Class Youth in the Northern United States, 1800–1900," by E. Anthony Rotundo in the *Journal of Social History,* Autumn 1989; as well as "When He Became My Enemy: Emerson and Thoreau, 1848–49," by Robert Sattelmeyer, published in the *New England Quarterly,* June 1989. Henry

Abelove's essay "From Thoreau to Queer Politics" is included in his book *Deep Gossip,* which takes its title from an elegy Allen Ginsberg wrote on the death of Frank O'Hara, "City Midnight Junk Strains," which mentioned O'Hara's ear as "a common ear," "for our deep gossip." In *Emerson,* Lawrence Buell makes the point that the Emerson-influencing-Thoreau model matches Victorian sensibilities—the father passing on his understanding to the son, in other words—whereas the Thoreau-usurping-Emerson model is a more modern narrative that suits an era that likes to think of itself as experimental even if it isn't. Lance Newman—in "Thoreau's Natural Community and Utopian Socialism," published in *American Literature,* September 2003—notes that Emerson didn't comprehend how social Thoreau was, as he criticized Thoreau for calling forests domestic, villagers urban, wilderness like Rome and Paris. "Of course this is exactly Thoreau's point," Newman writes. "What Emerson does not see is that, like *Walden,* 'A Winter Walk' is about having feelings of sociability, community, and even intimacy with wild nature itself—so much so that Thoreau's definiteness on the subject verges on redundancy." When I think of how I think Thoreau thought about sociability, I think of the theologian Karl Rahner, who wrote, "Our experience of ourselves occurs in unity with the experience of others."

As far as Emerson's domestic-servant experiment went, it should be noted that waged servants were free to move from domestic situation to domestic situation at the time, and that this too upset employers, their homes being disrupted in the process. On this topic, I relied on the work of the historian Barbara Ryan, who, in an essay entitled "Emerson's 'Domestic and Social Experiments': Service, Slavery, and the Unhired Man," wrote, "Emerson, in short, had come to see that payment of wages helped maintain hierarchies pertinent to privacy, mastery, and publication rights; that wages enforced distance, by insisting on the relative power of server and served; and that, at least for a capitalist, wages relieved guilt." Details about the state of the American economy in general came from the just-mentioned Lance Newman and from "The Social History of an American Depression, 1837–1843," by Samuel Rezneck, published in the *American Historical Review,* in July 1935. In discussing work and changing values in society, I also referred to *Disorderly Conduct: Visions of Gender in Victorian America,* by Carroll Smith-Rosenberg, as well as *What Hath God Wrought: The Transformation of America, 1815–1848,* by Daniel Walker Howe.

Thoreau's journal entry on fulfilling the promise of a friend's life is noted by Fink in *Prophet in the Marketplace.* I learned about American publishing in Gross's "Building a National Literature: The United States, 1800–1890," in *A Companion to the History of the Book,* and about the songwriter Francis Henry Brown in "Francis Henry Brown, 1818–1891, American Teacher and Composer," by Arlan R. Coolidge, in *Journal of Research in Music Education,* Spring 1961. My reading of the essays Thoreau wrote while in New York is partly informed by Steven Fink's reading in *Prophet* and by Howarth in *The Book of Concord.* Edward Hoagland's essay on Thoreau's mountain hikes is in *Elevating Ourselves: Thoreau on Mountains.* Philip Lopate's collection of various writers' New York writings, *Writing New York,* includes several long letters from Thoreau; Lopate notes, "Thoreau exemplified Walter Benjamin's remark that the nineteenth-century writer came to the city market ostensibly to observe, but actually to sell his wares."

I read about Fourierism in *The Utopian Vision of Charles Fourier: Selected Texts on Work, Love and Passionate Attraction,* which was translated and edited by Jonathan Beecher and Richard Bienvenu, and I used information from a review entitled "Association Forever: A New Look at the Fourier Movement," by Paul Boyer, in the March 1992 *Reviews in American History,* which reviewed *The Utopian Alternative: Fourierism in Nineteenth-Century America,* by Carl J. Guarneri. Information about the labor situation came from "Thoreau's Urban Imagination," by Robert Fanuzzi, published in the June 1996 issue of *American Literature*, as well as other sources. That Andrew Jackson called out the troops against workers on the canal came from a Historic Resource Study entitled *Chesapeake & Ohio Canal,* written by Harlan D. Unrau and published by the National Park Service. I read about women's labor in "The Historical Development of Women's Work in the United States," by Helen L. Sumner, in *Proceedings of the Academy of Political Science,* published in May 1971. The source of the song in which a mill worker says she does not want to die is a reminiscence by Harriet Hanson Robinson entitled *Loom and Spindle; or Life Among the Early Mill Girls*, published in 1898.

The idea of anyone dealing with manure, by the way, was new in Thoreau's time, a result of overworked farms in New England and the growth of cities, as noted in "Composting and the Roots of Sustainable Agriculture," an article by Barton Blum in *Agricultural History,* Spring 1992: "The demand for hay, the principal energy source for urban transportation systems, created even more dramatic demands for manure. Recycling was expanded by carting hay and other produce to cities, and returning to farm the manure produced in the consumption of this produce. The use of other urban organic wastes was also explored, including blood and offal from slaughterhouses; hair, flesh scrapings, and spent tanbark from tanneries; spent bone charcoal from sugar refineries; distillery refuse; the waste of glue factories; cottonseed; street sweepings; bones; sawdust; lime from paper mills; and waste from soap boilers and fish markets."

Information about Thoreau's pencil making and general engineering skills came from *The Pencil: A History of Design and Circumstance,* by Henry Petrowski, as well as from Edward Emerson's aforementioned book, and from an article entitled "Machine in the Wetland: Re-imagining Thoreau's Plumbago-Grinder," published in the *Thoreau Society Bulletin,* Fall 2005, but also available on the Thoreau Reader Web site (http://thoreau.eserver.org/), which is supported by the Thoreau Society and Iowa State University, and was also a great resource for this book. In writing about *A Week on the Concord and Merrimack Rivers,* I referred to the earlier-mentioned works of Fink and Howarth, as well as "A More Conscious Silence: Friendship and Language in Thoreau's *Week,*" by David Suchoff, in *English Literary History,* Autumn 1982. I also read "Common Places: Poetry, Illocality, and Temporal Dislocation in Thoreau's *A Week on the Concord and Merrimack Rivers,*" by Meredith L. McGill, in *American Literary History,* Summer 2007, which says of Walden: "As if speaking across the ages to the many Americans who annually make pilgrimages to Walden Pond in search of what, exactly, I'm not sure, Thoreau attacks that 'pathetic inquiry among travelers and geographers after the site of Ancient Troy. It is not where they think it is'. . . . Travelers cannot find Troy—or Walden, for that matter—for it is contemporary and coincident with the act of reading." John McPhee retraces the Thoreau brothers' boat ride in a story entitled "Five Days on the Concord and Merrimack Rivers," included in the collection *Uncommon Carriers;* a stand-in for one of the Thoreau brothers on McPhee's trip is Mark Svenvold, whose poems in the collection *Empire Burlesque* are very Thoreauvian in their mingling of long-ago faraway exploration—the Lewis and Clark Expedition—with the potluck of today.

In the chapters concerning the house literature in the time of Thoreau, I relied heavily on the writing of W. Barksdale Maynard, the architectural historian. I also relied on Robert Gross, who wrote: "Henry Thoreau stands out for the rigor of his social analysis. In the nineteenth-century dialectic of change, as he saw it, an *ancien régime* of traditional hierarchy and communal constraint was giving way to a new order of market domination and middle-class conformity. In the name of independence he repudiated both." For understanding "Civil Disobedience," I relied on Gross's "Quiet War with the State: Henry David Thoreau and Civil Disobedience," in the *Yale Review,* October 2005. I also relied heavily on an essay by Lawrence Rosenwald, professor of English and of Peace and Justice Studies at Wellesley College, in *A Historical Guide to Henry David Thoreau,* "The Theory, Practice and Influence of Thoreau's Civil Disobedience." A book that describes the simultaneous practical simplicity and radicalness of what "Civil Disobedience" is talking about is *Freedom Riders,* by Eric Etheridge, which contains portraits of the Freedom Riders today, accompanied by mug shots of them when they were jailed for riding a bus in Mississippi at the beginning of the civil rights movement, a collection of practical philosophers: "I can't speak for nobody else," one Freedom Rider recalls. "I didn't feel like I was a hero or anything like that. I did it and when I stopped doing it, I didn't feel like anybody needed to reward me or congratulate me or pat me on my back. I did what I felt like I had to do." In referring to a system that does not condone criticism, in reference to Abu Ghraib, I was referencing an idea presented in a talk I attended that was given by Philip Gourevitch, an author, with Errol Morris, of the book *Standard Operating Procedure.*

Other writings I relied on in the house-making chapter were "Thoreau's Enterprise of Self-Culture in a Culture of Enterprise," by Leonard N. Neufeldt, in *American Quarterly,* Summer 1987; "Picturesque Pattern Books and Pre-Victorian Designers," by Michael McMordie, in *Architectural History* (vol. 18), 1975; and "Pattern Book Parody in Walden," by Jean Carwile Masteller and Richard N. Masteller, in *The New England Quarterly,* December 1984. The Mastellers wrote, "Thoreau uses his imagination to turn pretense against itself as he insists that his fellow citizens can do better by turning their attention from housing sites to human insights." I read "Thoreau's First Year at Walden in Fact & Fiction," an address by Richard Smith that is published on the Thoreau Reader Web site. "Thoreau's House at Walden," by W. Barksdale Maynard, in *The Art Bulletin,* June 1999, which I also read, says: "To see Thoreau's sojourn in the context of the retirement phenomenon helps resolve a number of problems that have long troubled readers of *Walden,* including the apparent hypocrisy of the 'solitary' author's frequent visits to town. In the course of retirement—always a genteel habit—one was expected to maintain close ties with friends and relatives." An essential Thoreau book is Maynard's *Walden Pond: A History,* where I learned, for instance, that Van Wyck Brooks, who helped Thoreau enter America's literary pantheon in the early twentieth century, was turned on to Thoreau by George William Russell, widely known as AE, whose résumé has a Thoreauvian mix of occupations that includes mystic, farm policy theorist, and poet, and whose gravestone reads: "I moved among men and places, and in living I learned the truth at last. I know I am a spirit, and that I went forth in old time from the self-ancestral to labours yet unaccomplished."

From "Re-Creating Walden: Thoreau's Economy of Work and Play," by William Gleason, published in *American Literature,* December 1993, I learned about Catherine Beecher's *A Treatise on Domestic Economy.* Gleason notes that the writer David Reynolds, in *Beneath the American Renaissance,* has shrewdly argued, "Thoreau 'became the most compelling reform writer of nineteenth-century America' precisely because he 'recognized both the promise and the perils of contemporary reform movements.'" Gleason points out Thoreau's etymology of *labor* and writes about the influx of Irish into New England and their effect on the labor market and labor in general. He notes that Thoreau's ideas about the Irish shifted as Thoreau wrote the last drafts of *Walden,* but that his failure to allow the Irish into *Walden* as full citizens is a failure of *Walden.* I also read "Thoreau's Bog People," by Helen Lojek, in the *New England Quarterly,* June 1994.

For an understanding of Thoreau's lecturing skills, I read the writing of Bradley Dean, the much-loved Thoreau scholar, who died suddenly in 2006. "I owed Thoreau a big one," Mr. Dean was quoted as saying, in his obituary in the *Boston Globe*—the ponytailed Dean said Thoreau's writing had given him strength while serving time in a Navy brig in 1974, after refusing to get a haircut. In 1998, Dean was appointed media director of the Thoreau Institute by Don Henley, of the Eagles, who founded the nonprofit Thoreau study center. Before Dean died, Dean published Thoreau's last manuscript, *Wild Fruits,* which is the meditative and almost religious look at the berries and fruits of Concord, as well as a collection of letters from Thoreau to H. G. O. Blake, *Letters to a Spiritual Seeker*. At the time of the memorials for Dean, a member of the Thoreau Society was quoted as saying: "He never got to the point where he felt like he knew everything. It was always that process of discovery . . . that real intense passion that drove him."

I also read "Thoreau as Lecturer," by Hubert H. Hoeltje, in the *New England Quarterly,* December 1946, and Walter Harding's article "Thoreau on the Lecture Platform," in the *New England Quarterly* of September 1951. Thinking of Thoreau as a huckster and Yankee comic, I read *American Humor: A Study of the National Character,* by Constance Rourke, as well as "Carnival Rhetoric and Extra-Vagrance in Thoreau's *Walden,*" by Malini Schueller, in *American Literature,* March 1986. Thoreau's buffoon quote was brought to my attention in a review of his poetry in a 1965 article in *Modern Philology* by Charles Gruenert, then a professor at Eastern Montana College, now called Montana State University, Billings.

The relationship between the Brook Farmers and Thoreau was revised after the discovery of his visit to the farm by Sterling Delano, an expert on Brook Farm whose talk to the West Roxbury Historical Society was covered in the *West Roxbury Transcript* in the summer of 2008. The *Transcript*'s article quoted the West Roxbury Historical Society president as saying that we ought to look at where the farmers went post-farm. "I like to think of it as the 1840s hippies," he said. "It's the people that come out of Brook Farm—people early in the women's movement, early in anti-slavery movement—when you look at [them] after they leave, where they went in society, that's interesting." Delano said he wished there were more documentation of the nonliterary celebrities' lives. "I regret that I couldn't document the lives of people like you and me who were there," he said. "Good, salt-of-the-earth people who worked hard and lived at Brook Farm."

For the idea of Thoreau as an architect of an imaginary city, I am indebted to the aforementioned "Thoreau's Urban Imagination," by Robert Fanuzzi, an article that came to me when I thought I was insane to think that Thoreau had an urban imagination, a gift; it was published in *American Literature,* in June 1996. "A second look at Walden," says Fanuzzi, "suggests that Thoreau went to the country to find the city." He credits Thoreau with developing more than just a place: "What emerged at Walden Pond was not just an imaginary city but something even more incongruous with the American pastoral tradition—an urban imagination," Fanuzzi says. I learned about the etymological workings of the sandbank portion of *Walden* in Philip Gura's book *The Wisdom of Words: Language, Theology, and Literature in the New England Renaissance.* The lecture that compares *Walden* to a striptease is by Professor Ann M. Woodlief at Virginia Commonwealth University, and can be heard here: http://www.vcu.edu/engweb/audio/walden.html.

The Robert Frost definition of a poem—"Read it a hundred times and it will forever keep its freshness as a metal keeps a fragrance"—is mentioned in *The Winged Life: The Poetic Voice of Henry David Thoreau,* a collection of Thoreau's writings with excellent annotations by the poet Robert Bly, a poet that my editor, Gillian Blake, thinks is terrific. A view of *Walden* as an even more severe criticism of democracy is in "Thoreau's Critique of Democracy," by Leigh

Kathryn Jenco, in *The Review of Politics,* Summer 2003, where she writes, "Thoreau's own explanations for these acts reveal a much deeper concern that the theory and practice of democracy itself, not just democracy in its current manifestation, threaten the commitments that facilitate moral practice in our personal lives. By pointing out that such a system renders our voluntary responsibilities to ourselves and to our neighbors less compelling and meaningful, Thoreau indicts democracy for incurring real costs that, tragically perhaps, cannot be resolved by the system that created them." The etymology of the word *slum* comes from *How the Irish Invented Slang: The Secret Language of the Crossroads,* by Daniel Cassidy.

In looking at *Cape Cod,* I was inspired by Howarth's argument that in the book Thoreau "at last ponders the full meaning of the discovery of America." In a review of Howarth's book in the September 1982 *New England Quarterly,* Philip Gura notes: "When, for example, he focuses on Thoreau's search for the roots of American civilization on the sands of the outer Cape and reminds us that there Thoreau discovered how 'death is the dynamic principle in history' and so shapes 'the very movement of time—even as the Pilgrims and their Yankee descendants used their millennial theology in an attempt to deny the inevitable'—we understand how seriously Thoreau intended us to take his ocean book." I read "Thoreau's *Cape Cod:* The Unsettling Art of the Wrecker," by John Lowney, in *American Literature,* June 1992. In pondering Thoreau's thinking about death, I also read "'I heard a very loud sound': Thoreau Processes the Spectacle of Sudden, Violent Death," by Randall Conrad, in the *American Transcendental Quarterly,* June 1, 2005; it was originally delivered as a lecture at the 2002 American Literature Association convention, in which Thoreau Society panelists, as Conrad said, "had been asked to consider how (or whether) the Transcendentalists' philosophy can help twenty-first century citizens cope with a disaster of the magnitude of September 11, 2001." Jack Morgan's essay "Thoreau's 'The Shipwreck' (1855): Famine Narratives and the Female Embodiment of Catastrophe"—published in *New Hibernia Review* (Iris Éireannach Nua), in Winter 2004—notes, "The medieval horrors of the Famine were deeply in conflict with the rugged excitements of a youthful American republic. . . . The Famine, and the emigration experience that was an extension of it, were thus never admitted through the front door into American literature; rather, when the tragedy can be found at all in American letters, it takes place on the fringes." Morgan argues that the description of the dead woman's naked body "bespeaks political breakdown, the anti-structure that informs elemental horror." He relates the image to something out of the work of Stephen King.

I also read "Thoreau and the Irish," by Frank Buckley in the *New England Quarterly*, September 1940. Regarding the treatment of Irishwomen in particular, in his essay Peter Quinn sees disdain in Thoreau's description of the Irish famine victim; the essay was originally published in *American Heritage* under the title "The Tragedy of Bridget Such-a-One," though when including the essay in his collection *Looking for Jimmy: A Search for Irish America,* he changed the title to "The Triumph of Bridget Such-a-One." In explaining this revision, Quinn wrote, "Though it contained a multitude of tragedies, the Famine immigration to America eventually ended in triumph for the generation that followed." Quinn also quotes Yeats: "I dreamed that one had died in a strange place/ Near no accustomed hand;/ And they had nailed the boards above her face . . . / And left her to the indifferent stars above." Margaret Fuller's writing on the Irish in America, and Irishwomen in particular, is in *Woman in the Nineteenth Century, and Kindred Papers Relating to the Sphere, Condition and Duties of Women,* by Margaret Fuller Ossili, published in 1893. For more information on nativism, I read "The Amesbury-Salisbury Strike and the Social Origins of Political Nativism in Antebellum Massachusetts," by Mark Voss-Hubbard, in the *Journal of Social History,* Spring 1996.

Thoreau and Darwin's relationship is explored in the previously mentioned works of Richardson and Howarth.

For the look at Thoreau and the so-called cultural landscape, I relied on the work of David Foster, whom I visited at Harvard University's Experimental Forest, a 3,000-acre forest classroom and laboratory about an hour west on Route 2, in Petersham, Massachusetts. "I made a chart that showed the history of deforestation in New England," Foster told me as we hiked through the forest, the crackle of twigs accompanying him, "and Henry Thoreau's time is at the very peak of deforestation in New England. And here's the guy who went back to nature! He's our best nature writer and he's talking about wildness." Foster has read Thoreau's journals more times than I can imagine, and to walk through the woods with Foster, a biologist and ecologist, is to get an insight as to how Thoreau would have thought about his surroundings—and to how we have things reversed, as far as our popular conception of Thoreau goes. "When he considered the forested landscape around him," writes Foster, "Thoreau clearly viewed people as an inherent, though occasionally abusive, part of the natural world." (Such a woods walk is also a good time, owing to Foster's wit, which also crackles.) Aside from reading Foster's *Thoreau's Country,* I read *Forests in Time: The Environmental Consequences of 1,000 Years of Change in New England,* which Foster cowrote with John D. Aber. In the buildings at the Harvard Forest, I visited the forest dioramas, handcrafted depictions of the changes in the New England landscape from pre-colonial settlement times to today. They are as ornate and meticulous as they are beautiful and worth driving to, and they make computer imaging seem like an immature technology.

My understanding of Thoreau's so-called "reform papers," his political essays, comes from Hyde, Howarth, and Richarsdon, as well as from Deak Naber's book, *Victory of Law: The Fourteenth Amendment, the Civil War, and American Literature, 1852–1867.* Naber has written about the ways in which the literature of Thoreau and Thoreau's time shaped the Fourteenth Amendment, like DNA. It's an amazing story of how art, especially literature, informed the Fourteenth Amendment, which itself almost poetically reimagined the American concept of American justice, the way *Walden* hopes to reimagine the society. I also read " 'Hardly the voice of the same man': 'Civil Disobedience' and Thoreau's Response to John Brown," by J. J. Donahue, in the *Midwest Quarterly,* 2007 (vol. 48, no. 2).

In writing about "Wild Apples," I referred to Steven Fink in "The Language of Prophecy: Thoreau's 'Wild Apples,' " in the *New England Quarterly,* June 1986. "The essay deserves careful study if for no other reason than to dispel the impression that Thoreau's last years were ones of failure and decay," wrote Fink. Howarth's writing too has shown that Thoreau wasn't just sitting around after *Walden,* but was producing drafts and drafts of huge projects, as well as small ones.

I read about America's fascination with hermits in *Restless Souls: The Making of American Spirituality,* by Leigh Eric Schmidt. Roberts Haas's genetics of Thoreau's influence came from an article in the *San Francisco Bay Guardian,* "The Former Poet Laureate Keeps Spreading the Word," April 11, 2007.

The fact that the Indians that Thoreau visited would undertake an armed rebellion the following year on the lower Sioux reservation at Redwood Falls, Minnesota, comes from an article I read that described a study of the Minnesota Uprising by Dave Solheim, Dakota State University professor of English. After the uprising, the Indians were moved from Minnesota to the Dakotas and South Dakota. Solheim, a one-time associate poet laureate of North Dakota, has visited all the sites in Minnesota that Thoreau visited.

Thoreau's death is listed in the *New York Times,* on September 30, 1862.

You could write a whole book on Thoreau and nature, and how America's idea of nature has changed as the culture has changed and vice versa, and Lawrence Buell has done that with *The Environmental Imagination.* I relied on Buell's book especially in my last chapter in which I chronicle Thoreau's reputation and public interest in Walden Pond. I also read "English and American Criticism of Thoreau," by James Playsted Wood, in the *New England Quarterly,* December 1933; "Henry S. Salt, the Late Victorian Socialists, and Thoreau," by George Hendrick, in the *New England Quarterly,* September 1977; and, as previously noted, "Towards the Making of Thoreau's Modern Reputation." For Thoreau's political reputation, I relied on *Several More Lives to Live: Thoreau's Political Reputation in America,* by Michael Meyer. I also read *Thoreau in the Human Community,* by Mary Elkins Moller.

Discussing Route 2, I referred to Richard Forman's article "The Ecological Road-Effect Zone of a Massachusetts (U.S.A.) Suburban Highway," in *Conservation Biology,* February 2000. (I intended to visit Forman at Harvard but didn't because I couldn't find any parking, which serves me right.) I spoke with Richard Primack on my cell phone while in a park in Red Hook, Brooklyn. I had learned of Primack's work from a *Boston Globe* article in which he wrote about New England climate conditions, where in Concord spring now comes about a week earlier than in Thoreau's time: "We hear about the effects of global climate change on hurricane systems, Gulf Streams, the melting of glaciers, but a lot of this information seems very far away," Primack has said. "It seems like it's happening in some other part of the world. But we want to use Massachusetts—and Concord, in particular—as a case study which demonstrates that global warming is happening and it can't be ignored."

Remarks by Whitman are taken from *Walt Whitman: His Life and Work,* by Bliss Perry, and from *With Walt Whitman in Camden,* by Horace Traubel, published in 1914.

When I mention Thoreau's prophetic voice, I am doing so as a result of having read Lewis Hyde's *The Gift,* as well as his essays on Thoreau: "Rather," Hyde has written, "the prophet speaks of things that will be true in the future because they are true in all time."

Richardson said that Emerson wrote the best essay on Thoreau. In this century, what with our Thoreau complications, I would say one of the best essays is Rebecca Solnit's "The Thoreau Problem," first published in *Orion* and now the final piece in a collection of environmental writing assembled by Bill McKibben, *American Earth,* which is where I read it. She discusses the misunderstanding that idealists like Thoreau do not take pleasure in things—that they are aesthetes, humbugs, and worse. She writes: "This schism creates, as the alternative to a life of selfless devotion, a life of flight from engagement, which seems to be one of the ways those years at Walden are portrayed. But change is not always a revolution, the deprived don't generally wish that the rest of us would join them in deprivation, and passion for justice and pleasure in small things is not incompatible. That's part of what the short jaunt from jail to hill says."

For Thoreau's relationship with music, I read "Thoreau: The Ear and the Music," by Kenneth W. Rhoads, in *American Literature,* November 1974, which describes so many of the times Thoreau would stop and listen to music being played, sung, or whistled all around him in the average nineteenth-century day. As far as my own musical life goes, I have lately heard my teenage son sing bits and pieces of "Tom Bowling," and my daughter plays the fiddle tune, "Highland Laddie," that Thoreau and his neighbors are said to have danced to; I have profited greatly from these things. A good recording of "Highland Laddie" is by Ned Pearson, an English fiddler, on *Ranting and Reeling: Dance Music of North England,* part of an anthology called Music of the People.

My favorite Brooklyn lilac bush is very near the East River in Vinegar Hill, across from an air-fouling power plant that I rely on every day, I would imagine. In deciding whether or not Whitman would have walked past that corner, I read a lot and talked to a lot of people, including my neighbor Michael Cassidy, a local photographer who has done a lot of reading and talking about the area himself. Neighbors of the lilacs told me that the bushes were only a few decades old. On the walk over to show the lilacs to my wife and daughter, we ran into a guy who said he had a house in Vermont and that when he comes upon lilac bushes, he knows he'll find the foundation of an old farmhouse, as lilacs indicate ruins. On the way home, I collapsed with a fever and went to sleep for many hours. I rarely get so sick. When I woke I felt as if I had died and been born again, or something like that. Anyway, I knew I had an ending to a long Thoreau book.

Index

Robert Sullivan is the author of *The Meadowlands*, *A Whale Hunt, How Not to Get Rich,* and the best sellers *Rats* and *Cross Country*. He is a contributing editor to *Vogue,* and his writing has appeared in many magazines and newspapers, such as the *New York Times*, the *Los Angeles Times,* the *Oregonian, The New Yorker, The New Republic, Dwell, Craft, Men's Vogue, Rolling Stone, Runner's World, Condé Nast Traveler,* and *Outside*. He lives in Brooklyn, New York, with his wife and two children.